Philosophy and Neuroscience

Studies in Brain and Mind

Volume 2

Series Editors

Philosophy and Neuroscience

A Ruthlessly Reductive Account

by

John Bickle
University of Cincinnati, U.S.A.

KLUWER ACADEMIC PUBLISHERS
DORDRECHT / BOSTON / LONDON

Library of Congress Cataloging-in-Publication Data

Bickle, John.
Philosophy and neuroscience : a ruthlessly reductive account / by John Bickle.
p. ; cm. -- (Studies in brain and mind ; 2)
Includes bibliographical references and index.
ISBN 1-4020-7394-1 (alk. paper)
1. Neurosciences--Philosophy. I. Title. II. Series.
[DNLM: 1. Psychophysiology. 2. Mind-Body Relations (Metaphysics) 3. Neuroscience.
4. Philosophy. 5. Psychological Theory. WL 103 B583p 2003]
RC343.B43 2003
612.8'01--dc21

2003040140

ISBN 1-4020-7394-1
ISBN 1-4020-1302-7

Published by Kluwer Academic Publishers,
P.O. Box 17, 3300 AA Dordrecht, The Netherlands.

Sold and distributed in North, Central and South America
by Kluwer Academic Publishers,
101 Philip Drive, Norwell, MA 02061, U.S.A.

In all other countries, sold and distributed
by Kluwer Academic Publishers,
P.O. Box 322, 3300 AH Dordrecht, The Netherlands.

Printed on acid-free paper

Printed in the Netherlands.

To Marica

To Caroline, Kat, and Margaret

Family

"And yet, there are philosophers who refuse to acknowledge scientific philosophy as a philosophy, who wish to incorporate its results into an introductory chapter of science and claim that there exists an independent philosophy, which has no concern with scientific research and has direct access to truth. Such claims, I think, reveal a lack of critical judgment. Those who do not see the errors of traditional philosophy do not want to renounce its methods or results and prefer to go on along a path which scientific philosophy has abandoned. They reserve the name of philosophy for their fallacious attempts at a superscientific knowledge and refuse to accept as philosophical a method of analysis designed after the patterns of scientific inquiry.

What is required for a scientific philosophy is a reorientation of philosophic desires."

--Hans Reichenbach (1957), *The Rise of Scientific Philosophy*, 305.

CONTENTS

LIST OF FIGURES

PREFACE

As a work that combines philosophy and (what was then) current science, I have always admired David Hull's (1974) *Philosophy of Biological Sciences* in the Prentice Hall "Foundations of Philosophy Series." Hull's writing is crisp, his presentation is focused, and—most importantly—the scientific details, clearly described, drive the philosophy. I kept these features of Hull's book constantly before me as I wrote this manuscript. I hope others see in this work those same features, along with the addition of more science from the primary literature, as befitting a book that seeks to do more than introduce students to a philosophical area.

Neuroscience and its implications have not gone unnoticed by either professional philosophers or the educated general public. How could they have? We move closer every day to actually having something that human beings have speculated about for centuries, a purely physical account of behavioral causes. Yet it has struck me for close to a decade that most self-described "philosophers of neuroscience" and "neurophilosophers"—people with a professional stake in keeping up with the actual science—focus on the wrong levels of research, theory, and experiment. Furthermore, this mistaken emphasis by specialists produces negative repercussions in the larger philosophy of mind/cognitive science community. Without question, neural network modeling and computer simulation, functional neuroimaging, and neuropsychological and neurological assessments are central neuroscientific pursuits. But one glance at an influential neuroscience textbook, or a short perusal through recent titles of colloquia talks delivered at a Ph.D.-granting Neuroscience department, or—even more dramatically—one visit to the week-long Society for Neuroscience annual meeting, should convince anyone that the amount of research being done in "cognitive neuroscience" and the amount we can safely be said to "know" at that level pales in comparison to the amount going on and already discovered in the discipline's cellular and molecular core. I'll begin developing this theme right off the bat in Chapter One, but this entire book is at bottom an extended argument that higher-level theorists of mind, especially philosophers, should reorient their interests "down levels" in the neurosciences. Or, short of that, they should realize that the mainstream core of the current science, the part on which *all* the higher level cognitive neuroscientific investigations ultimately depend, has a "ruthless" reductionism built directly into its practice. Furthermore, at this cellular/molecular level, we "know a lot about how the brain works" and we are increasingly able to manipulate specific behaviors by intervening directly with these cellular processes and intracellular pathways. This is no longer just the fare of science fiction, but of *Cell*, *Journal of Neurophysiology*, *Science*, *Nature*, and *Proceedings of the National Academy of Sciences*.

The status of "ruthless reductionism" in current mainstream neuroscience contrasts sharply with its status in philosophy, even in the tip of the "analytic" branch that over the last century embraced scientific philosophy. In his Presidential Address to the American Philosophical Association (delivered orally in 1989), noted reductionist philosopher Jaegwon Kim remarked:

> Perhaps as a result of the singular lack of success with which our earlier reductionist efforts have been rewarded, a negative image seems to have emerged for reductionisms in general. Many of us have the feeling that there is something rigid and narrow-minded about reductionist strategies. Reductionisms, we tend to feel, attempt to impose on us a monolithic, strait-jacketed view of the subject matter, the kind of cleansed and tidy picture that appeals to those obsessed with orderliness and discipline. ... Perhaps, too, reductionists are out of step with the intellectual style of our times: we strive for patterns of life and thought that are rich in diversity and complexity and tolerant of disagreement and multiplicity. We are apt to think that the real world is a messy place and resists any simplistic drive, especially one carried on from the armchair, toward simplification and unification. In fact, the word "reductionism" seems by now to have acquired a negative, faintly disreputable flavor—at least in philosophy of mind. Being a reductionist is a bit like being a logical positivist or a member of the Old Left—an aura of doctrinaire naiveté hangs over him. (1993, 265-266)

Kim's assessment of philosophical orthodoxy remains correct to this day.

The motivation guiding this book is that this orthodoxy remains because (scientific) reduction is misunderstood. That diagnosis is, of course, not original with me. But my new prescription for the malady is. I now contend that the only way to overcome this misunderstanding is to show *what scientific reductionism is in practice*—the experiments it motivates, the results these experiments have yielded, and the way they are interpreted—in the mainstream branch of a "hot" reductionist discipline. I leave it to readers, both philosophers and scientists, to decide if this prescription works.

Realistically, however, can I expect philosophers and cognitive scientists to wade through as much cellular and molecular detail as I have included in this book? I hope so. There is a growing schism in both philosophy of mind and philosophy of science, between metaphysically minded and normatively prescriptive philosophers versus philosophers willing to countenance scientific practice and results as scientists present them. I'm trying to push the agenda of the second camp one step beyond where it has been pushed so far by philosophers interested in neuroscience. I seek to push it into *core* neuroscience circa 2002. If successful, this will widen philosophy's schism. But that might not be such a bad thing. Perhaps it is time to cleave philosophy of mind, philosophy of science, and philosophy of particular sciences (like psychology, cognitive science, neuroscience, and biology) into separate disciplines: one that, although mindful of scientific practices and results, remains tied to perennial metaphysics and epistemology; the other a part of science itself. This attitude reflects the rationale behind the Reichenbach quote that serves as this book's epigram, as well as the first part of my title. This is a book on philosophy *and* neuroscience, not philosophy *of* neuroscience. Does the neuroscience overshadow the philosophy? As readers will see throughout this book, I take on questions and arguments that have been put forward by philosophers; only I do so while limiting myself to the resources of recent cellular and molecular neuroscience. I also strive for the "synoptic vision" of all of neuroscience that lies implicit in its mainstream cellular and molecular core. In one sense I do leave things "entirely up to science," but in another I am putting together the individual pieces that science provides to make explicit the "bigger picture" that most scientists leave implicit. That's "philosophy" enough for me.

My subtitle comes from my colleague, Robert Richardson. Those who know Bob know of his ruthless wit. Appropriately, he heard my original, boring subtitle, *A Thoroughly Reductive Account*, and immediately suggested the much punchier form. Continuing discussions with a number of my colleagues in Philosophy at the University of Cincinnati—in particular, Richardson, Don Gustafson, Christopher Gauker, and Tom Polger—helped me clarify arguments (and strengthened my conviction that I was on the right track!). The Neuroscience Graduate Program at the University of Cincinnati College of Medicine, of which I am very proud to be a part, keeps me up on the latest scientific developments and trends through its weekly visiting speaker's seminar. My scientific collaborators on a recent functional neuroimaging project—Scott Holland at the University of Cincinnati and Childrens Hospital, Cincinnati, Malcolm Avison at the University of Kentucky Medical Center, and Vince Schmithorst at Childrens Hospital, Cincinnati—have all commented helpfully on these themes over many discussions. My current Ph.D. student, Anthony Landreth, assisted with this manuscript in numerous ways. I tried out earlier versions of this material in a Fall 2000 seminar at the University of Cincinnati, and thank Ph.D. students from Philosophy, Biology, and Neuroscience for helpful questions and comments.

I've presented talks that became pieces of this book at many conferences and colloquia over the past few years, so thanks to all who've made me think things through again with their questions and comments. This group includes especially Ken Aizawa (Centenary College), Louise Antony (Ohio State University), Ansgar Beckerman (Univresity of Bielefeld), Luc

Faucher (University of Quebec at Montreal), Owen Flanagan (Duke University), Nicolas Georgalis (East Carolina University), Melvin Goodale (University of Western Ontario), Valerie Hardcastle (Virginia Tech), Terry Horgan (University of Arizona), Huib Looren de Jong (Vrije University Amsterdam), Jaegwon Kim (Brown University), Michael Quante (University of Muenster), Maurice Schouten (University of Tilburg), Sven Walter (Ohio State University), and Ümit Yalçin (East Carolina University). A number of philosophers, neuroscientists, and psychologists commented on this manuscript's penultimate draft and numerous improvements resulted. This group includes Jim Bogen (University of Pittsburgh), Carl Craver (Wahsington University in St, Louis), Dingmar van Eck (Vrije University Amsterdam), Trent Jerde (University of Minnesota), Huib Looren de Jong (Vrije University Amsterdam), Maurice Schouten (University of Tilburg), John Symons (University of Texas at El Paso), and Kenneth Sufka (University of Mississippi). Thanks also to graphics artist David Winterhalter, who adapted and created some of the figures. A Publications grant from the University of Cincinnati Taft Memorial Fund supported production of the illustrations.

The biggest thanks of all, of course, go to Marica Bernstein and my family, Caroline, Kat, and Margaret Cooper. Marica not only created and adapted some of the figures (right before the deadline for the final manuscript!), but also helped me write some of the scientific sections covering contemporary molecular genetics and, as always, was a critical respondent to every idea presented here. Together we live the excitement of our shared scientific interests, and I can't imagine a better partner to share my professsional and personal lives. They, like Marica and I, meld into one.

November 2002
Cincinnati, Ohio

CHAPTER ONE
FROM NEW WAVE REDUCTION TO NEW WAVE METASCIENCE

This book is about contemporary neuroscience. More specifically, it works with detailed examples drawn from current research to express that discipline's reductive aspirations, aims, and potential. This reductionism holds important consequences for some "hot" issues in contemporary philosophy of mind. Even more specifically, this book is about the nature of reduction at work in the mainstream core of the current discipline, cellular and molecular neuroscience.

Questions arise immediately. Why look for philosophical lessons in current cellular and molecular neuroscience? Why does the nature of reduction at work in this area warrant interest, philosophical or scientific? And why try to provide this account by focusing on detailed examples of current research instead of, say, articulating a general account of scientific reduction and applying it to them? These are questions I'll tackle in this introductory chapter.

1 WHY CELLULAR AND MOLECULAR NEUROSCIENCE?

Some philosophers (of mind and science) and cognitive scientists regularly keep abreast of developments in contemporary neuroscience. Patricia Churchland's landmark *Neurophilosophy* (1986) provided an explicit defense of this interdisciplinary attention. In the book's "General Introduction," she writes: "In a way, nothing is more obvious than that philosophers of mind could profit from knowing at least something of what there is to know about how the brain works. After all, one might say, how could the empirical facts about the nervous system fail to be relevant to studies in the philosophy of mind?" (1986, 4). Even philosophers who don't share Churchland's exuberance for neuroscience can agree with this much. It only requires that they take discoveries in our current sciences as relevant for some philosophical issues. Of course, there are philosophers who reject even this. They are beyond the pale (of this book).

However, neuroscientifically astute philosophers and cognitive scientists have almost universally ignored the "cellular and molecular wave" that swept through neuroscience over the past two decades. Instead, they've focused on "cognitive neuroscience," the "interdisciplinary melding of studies of the brain, of behavior and cognition, and of computational systems that have properties of the brain and that can produce behavior and cognition" (Kosslyn 1998, 158). Investigative techniques here range from state-of-the-art functional neuroimaging to traditional neuropsychological measures to computational modeling in massively interconnected neural networks. It isn't surprising that philosophers (and cognitive scientists) with neuroscientific proclivities are attracted to this branch of the discipline. First, there is the nearly universal intuition among high-level theorists that "levels" considerations and relations are crucial to understanding the mind-brain. In fact, the mind-brain need not be thought of as special in this regard. Most philosophers of biology assume a similar view about the importance of levels in the study of higher-level biological phenomena; the "philosophy of molecular biology" is hardly a recognized area. In addition to these "levels" intuitions, there is also familiarity. Philosophers are at home with cognitive neuroscience's descriptions of behavior and cognition, and with the types of behavior and cognition these scientists investigate. Grasping the experimental methods isn't even much of a professional stretch. The physics of functional neuroimaging are daunting, but even practicing cognitive neuroscientists who employ these methods tend to leave their physics to physicists and concentrate on the behavioral and control tasks and the functional interpretation of analyzed data. The basic concepts of neurocomputational techniques and their mathematics are readily presented geometrically, making them comprehensible to anyone with some quantitative background (Churchland and Sejnowski 1992). Finally, the relevance of cognitive neuroscientific theories and explanations for philosophical (and cognitive scientific) issues is usually readily apparent. Many philosophers are interested in "naturalizing" intentionality, consciousness, and the like. Cognitive neuroscience appears to be a direct scientific analog of philosophical "naturalizing" projects. In short, the levels of theory and explanation inhabited by contemporary cognitive neuroscience are nearby those of scientifically inspired philosophy of mind. So why search through other branches of current neuroscience for philosophical consequences and implications?

The principal reason is straightforward: neuroscience's "mainstream" currently lies elsewhere. It lies in cellular physiology and molecular biology. This "revolution" began two decades ago and now is in full swing. It is in keeping with the ascendance of molecular techniques and investigations in biology generally and is now reflected clearly in principal neuroscience textbooks. Consider a single example. A decade ago, in the introduction to the third edition of their monumental *Principles of Neural Science*, Eric Kandel, James Schwartz, and Thomas Jessell asserted the promise of investigating the molecular mechanisms of mind: "The goal of neural science is to understand the mind, how we perceive, move, think, and remember. In the previous editions of this book we stressed that important aspects of behavior could be explained at the level of individual nerve cells. ... Now it is possible to address these questions directly on the molecular level" (1991, xii). Do notice the first sentence. The ultimate *explanandum* of neuroscience is mind, not some behavioral or ersatz laboratory substitute. By 1991, the search was already on for its molecular mechanisms and their experimental verification, to the extent that this focus had already made the discipline's general textbooks.

By the text's recent fourth edition, and after another decade of cellular and molecular research, these same authors were ready to announce mind-to-molecules "linkages" not just as research promises, but rather as accomplished results:

> This book ... describes how neural science is attempting to link molecules to mind—how proteins responsible for the activities of individual nerve cells are related to the complexity of neural processes. Today it is possible to link the molecular dynamics of individual nerve cells to representations of perceptual and motor acts in the brain and to relate these internal mechanisms to observable behavior. (2000, 3-4)

The chain of explanations envisioned by these authors is nothing less than a *reduction* of mind to molecules, through interposed "cognitive" and cellular levels. It should be noted explicitly that *Principles of Neural Science* remains the standard comprehensive textbook in the field.

Two lessons from these passages—and from similar passages that occur in introductory chapters in most of neuroscience's current texts—are crucial. First, according to these prominent neuroscientists speaking with the authority of textbook authors, some observable behaviors have already been explained at the level of molecular mechanisms. Second, the guiding aim of "mainstream" neuroscience is the discovery of these mind-to-molecules "linkages." So by limiting attention to cognitive neuroscience only, by ignoring the cellular and molecular core, philosophers and cognitive scientists are getting off the neuroscience train before the current end of its explanatory line. Throughout this book I will argue explicitly that techniques of cognitive neuroscience are an essential part of discovering mind-to-molecules "linkages." But some of what these techniques reveal has already been carried "further down" to cellular, synaptic, and ultimately molecular biological mechanisms; these existing reductions reveal an essentially heuristic role for higher-level scientific investigations; and there is genuine empirical promise that more "ruthless reductions" will be coming forth. Defending these assertions is this book's principal goal.

When one limits his or her attention to cognitive neuroscience, one not only misses some of neuroscience's most celebrated recent results. One also misses the core of the current discipline: the problems, methods, and results that occupy the day-to-day work of the greatest percentage of the 28,000+ scientists who belong to the Society for Neuroscience and thereby identify themselves professionally as neuroscientists. The Society is "the largest professional society dedicated to the study of the nervous system." Regular membership is open to "any scientist ... who has done meritorious research relating to the nervous system" (www.sfn.org/memb/fact_sheet.html). Its web site offers a searchable database of the abstracts from the 13,000+ slide talks and posters presented at the most recent annual meeting (at this writing, of the 2001 annual meeting, November 4-11, in San Diego, CA) (http://sfn.scholarone.com/, available as a link from www.sfn.org). This resource is a gold mine for surveying "hot" areas in current neuroscience. What is hot now? Searching for cellular and molecular mechanisms and attempting to relate these mechanisms directly to observable behavior. For example, if one conducts a standard search of this database on the two themes most dominated by cellular physiological and molecular biological experimental techniques, namely "Development" and "Synaptic Transmission and Excitability," searching for "Any subtheme," one receives 4698 abstracts (respectively, 1818 and 2880). On the other hand, if one searches the theme most closely associated with cognitive neuroscientific techniques, namely "Cognition and Behavior," one receives 1873 abstracts. But when one further limits this last category to the subtheme, "Human cognition and behavior," the number drops to 476 abstracts; and many of these are purely behavioral studies that don't purport to offer "cognitive" explanations for the data revealed. This is anecdotal evidence, of course, but the upshot is clear. When it comes to basic scientific, not-purely-clinical, research, the search for cellular and molecular mechanisms dominates among bench neuroscientists.

Amateur sociology aside, there is another way that philosophers and cognitive scientists become misinformed when they limit their attention to cognitive neuroscience. Searching for mind-brain connections at the level of regional neural activation patterns, neuropsychological measures, or activation vectors across neural networks isn't as "ruthlessly reductionistic" as attempting to explain behavior—and manipulate it experimentally in animal preparations—at the level of intracellular signaling pathways within and between the individual neurons constituting the network. The "parts" are smaller in the latter theories and there is intuitively a larger gap between starting and ending "levels," between *explananda* and *explanans*. One will thereby misunderstand the nature and extent of reductionism in current neuroscience if one knows nothing about the research and theory now taking place to forge mind-to-molecules "linkages."

So this book's first task is to reveal the scientific details of some accomplished mind-to-molecules "linkages" and to evaluate the explanatory potential of this "ruthless reductionism" for behavior and cognition generally. Only by considering these details can one grasp the reductionism that characterizes the core branch of current neuroscience. One might end up disagreeing with my positive arguments for the explanatory potential of current and foreseeable reductionist cellular and molecular neuroscience. To tip my hand, I will argue for its explanatory potential all the way up to features of consciousness (in Chapter Four). But no one should be mistaken about the *factual existence* of a ruthless and audacious reductionism that informs neuroscience's current cutting edge. If I can communicate that, I will at least break the popular but mistaken myth among philosophers and cognitive scientists that reductionism is "dead." On the contrary: it is alive and thriving, at the very core of one of the hottest (and best funded) scientific disciplines. Perhaps I can even convince some that this "ruthless reductionism" is the *correct* way to pursue a science of mind, given all we know and can do now.

One final remark needs to be emphasized before we start doing philosophy and neuroscience in earnest. I've already referred to current cellular and molecular neuroscience as trendy, "hot," and "well-funded." It is easy to misunderstand the significance I place on these features. I am well aware that even many neuroscientists find these terms unseemly and are squeamish about the sociology of their discipline that these terms reflect. (I myself do not and am not, but that is a mere psychological fact about me that has no bearing on the arguments in this book.) I am also well aware that there are numerous (sociological) explanations for why this branch of the discipline is so prominent, including funding opportunities for expensive research equipment, novel (and very cool) technological toys to play around with, the larger *Zeitgeist* that propels molecular biology in general at present—add your personal favorite here. However, I am NOT committing a simple-minded fallacy that equates a discipline's temporal popularity and external funding trough with a beeline to Truth or the fruitfulness of its approach!

Instead, my appeal to the centrality of cellular and molecular neuroscience within the discipline serves two purposes in setting up this book's project (and in motivating particular steps in the argument later on). First, it stresses the fact that "naturalistic" philosophers of mind, committed to the continuity of philosophical theses and arguments with neuroscientific work (including some who describe their research as "philosophy of neuroscience") have been focusing myopically and virtually without exclusion on work that constitutes a tiny island of that discipline; and given the "higher level" location of that tiny island, these philosophers have not seen how ruthlessly reductive that the bulk of neuroscience actually is circa the turn of the 21st century. A central task of this book (beginning in Chapter Two) is to *demonstrate explicitly* this ruthless reductionism—rather than just assert it, as I am doing now. Second, even though I acknowledged multiple explanations for why an approach can become trendy, hot, and funding rich, I am also convinced that in this particular case the "on the right track" reason is among these. If you are convinced that the sociological factors exhaust the reasons for cellular and molecular neuroscience's "hot" and "well-funded" current status, then this book probably isn't for you. However, I am also well aware of a demand to *argue* for this thesis, and that will consist in explaining not just some results of current cellular and molecular neuroscience, but of the experimental methods, their rationale, and the case for their expected fruitfulness. It will also consist in showing how these scientific resources can be put together to address important philosophical concerns. In other words, the rest of this book constitutes an argument for the scientific and the philosophical legitimacy of neuroscience's cellular and molecular revolution. For those who are squeamish about terms like "trendy," "hot," and "well-funded," or for those who assume when they see such terms that a simple-minded fallacy about the justifiability of an approach is on offer, ear-marking this page might be in order, to remind themselves about the significance and the challenge to argument that I place on these features.

2 BACKGROUND: THE INTERTHEORETIC-REDUCTION REFORMULATION OF THE MIND BODY PROBLEM

Although the nature and extent of reduction in current cellular and molecular neuroscience is unfamiliar to most philosophers of mind (and cognitive scientists), the concept of scientific reduction is not. Theories of intertheoretic reduction from philosophy of science have been mobilized and criticized by philosophers of mind and psychology over the past four decades. The project I am pursuing in this book is a self-conscious attempt to break an impasse in current philosophy of mind over the importance of scientific reduction for addressing "the mind-body problem." To set up my project, I will first explain how philosophy of mind got into this impasse (an impasse which includes my previous "new wave reductionism" in Bickle 1998). Then

I'll suggest how my new approach and its larger philosophical motivations purport to break this impasse. Besides, this episode in recent philosophy is interesting enough to warrant presenting to this book's wider interdisciplinary audience.[1]

An old philosophical question about "the relationship between mind and body" lies at the bottom of any discussion of reduction in the psychological and brain sciences. Neuroscientists have not missed this foundation. Introductory chapters of general neuroscience textbooks routinely begin with a quick overview of philosophical views about mind, usually starting with René Descartes' interactive dualism of mental and physical substances. Unfortunately, many of these discussions depart from philosophy's history before the middle of the 20th century, by the time that substance dualism had lost almost all popularity with philosophers in the "analytic" tradition. From introductory chapters in neuroscience texts, one can obtain the mistaken impression that *materialist* or *physicalist* views about mind barely exist in philosophy.[2] Paul Churchland's masterful (1987) introductory philosophy of mind text is a welcome panacea. After presenting standard arguments against dualist views of mind, Churchland writes: "Arguments like these have moved most (but not all) of the professional community to embrace some form of materialism" (1987, p. 21). That "professional community" includes many philosophers in the "analytic" tradition.

How does an interest in scientific reduction fit into recent philosophy of mind? The traditional philosophical mind-body problem is *ontological*. "Ontology" is a piece of fancy Greek philosophical jargon for the study of being, of "what there is," of the fundamental constituents and categories of existence. The mind-body problem is about the fundamental nature of the conscious and cognitive mind and its relation to physical events. At bottom, it is a clash between two conflicting intuitions:

1. That the nature and core properties of mental phenomena guarantee that they cannot be identical to physical (i.e., neural) events;
2. That the domain of the mental should ultimately be brought within the scope of our otherwise comprehensive and wholly physical scientific worldview.

Physicalism amounts to giving pride of place to intuition #2.

But how can one *argue* for this view? How can one support such an account as more than "blind faith" or "mere opinion"? One way, fashionable for nearly four decades now, is to construe intuition #1 as resting on allegiance to a psychological theory whose central theoretical posits (beliefs and desires, or propositional attitudes generally) figure in generalizations that explain and predict observable behavior. The "nature and core properties of the mental," paraded as "conceptually autonomous" by defenders of intuition #1, are said to depend upon this theory's constitutive principles and generalizations. One can then reformulate the traditional philosophical mind-body problem as, first and foremost, a question about the intertheoretic relationships obtaining between this psychology and its counterparts or successors from the physical sciences (like neuroscience). Ontological conclusions become secondary to and dependent upon the nature of the cross-theory relations that obtain. The ontology of mind is thus treated on a par with similar cross-theoretic conclusions in science—like, e.g., the relationship between temperature and mean molecular kinetic energy. Historically in science, *intertheoretic reduction* has been the key relation thought to yield cross-theoretic conclusions about the entities and properties posited by potentially reducible theories. Depending upon whether or how an intertheoretic reduction goes, scientists draw conclusions ranging from

- the autonomy of a theory's objects and properties (e.g., electrical charge is not identical to any combination of an object's mechanical properties),
- cross-theoretic identity (e.g., visible light is electromagnetic radiation with wavelength between .35 and .75μm),
- conceptual revision (e.g., mass is a two-place relation between an object and countless reference frames, not a one-place property of objects), or
- elimination (e.g., there is no such thing as phlogiston).

It has also been fashionable among some physicalists to claim that something called "folk psychology" provides the theory with which our familiar ontology of mind is affiliated. Folk psychology is supposed to be the rough-and-ready patchwork of generalizations that underwrite our everyday explanations and predictions of each others' behavior. Its principal theoretical posits are beliefs and desires, which mediate between sensory inputs and behavioral outputs.[3] There is, however, no need to saddle our ontology of mind with "folk psychology." One can give our mentalistic categories and kinds all the resources and sophistication of scientific psychology and ask how those theories are relating to developing neuroscientific counterparts. Friends of the mental can help themselves to any psychological theory they see as fit to champion, be it commonsensical or sophisticatedly scientific.[4]

Call this approach the Intertheoretic Relation (IR) reformulation of the mind-body problem. Its guiding hope is that by reorienting the traditional philosophical issue away from its ontological focus, making ontological conclusions secondary to the prior intertheoretic reduction question, we might overcome the deadlock over intuitions #1 and #2 in a way that makes relevant some of the rich and rigorous resources of 20th century philosophy of science and the contemporary cognitive and brain sciences. Scientific clarity hopefully emerges from philosophical murk.

So it is not surprising that soon after Ernest Nagel (1961, chapter 11) published his groundbreaking theory of intertheoretic reduction, it was adopted by physicalist philosophers of mind. Physicalists especially liked his linking the "temperature-is-mean molecular kinetic energy" identity to the intertheoretic reduction of classical thermodynamics to statistical mechanics. Nagel's resources became so prevalent that critics took themselves to be attacking the entire psychoneural reductionist program by pointing out difficulties that "special sciences" (like economics and psychology) pose for Nagel's account. For example, in footnote 2 of his ([1974] 1981) Jerry Fodor asserted—without argument—that "many of the liberalized versions of reductionism suffer from the same base defect as what I shall take to be the classic form of the doctrine" ([1974] 1981, 322). The "classic form" was Nagel's account, which had been published only *thirteen years* prior to Fodor's essay. (Some works become "classics" very quickly!)

In the spirit of logical empiricism—the paradigm of mid-20th century philosophy of science to which he adhered and partly defined—Nagel's theory of intertheoretic reduction purported to be completely general. It purported to apply to every specific case in science and its history. It held that the reduction of one theory to another consists of a logical *deduction* (derivation) of the laws or principles of the reduced theory $\mathbf{T_R}$ from the laws or principles of the reducing or basic theory $\mathbf{T_B}$. In interesting cases ("heterogeneous cases," as Nagel called them), $\mathbf{T_B}$'s descriptive vocabulary lacks terms contained in $\mathbf{T_R}$. For example, "pressure" and "temperature" do not occur in statistical mechanics or microphysics. The logical derivation in such cases also requires various "correspondence rules" or "bridge principles" (**BP**) to connect the disparate vocabularies, at least one for each term of $\mathbf{T_R}$ not included in $\mathbf{T_B}$. (We'll hear more about **BP**s in the next section.) Eschewing niceties and many details, we can represent Nagel's account as follows:

$\mathbf{T_B}$ & **BP** (as necessary)
logically entails
$\mathbf{T_R}$.

Also in the spirit of logical empiricism, Nagel characterized $\mathbf{T_R}$, $\mathbf{T_B}$, and **BP** "syntactically," as sets of statements or propositions distinguished by their logical form.

Reformulated in light of Nagel's theory, the traditional philosophical mind-body problem becomes: Will future brain science develop theories (**T_B**s) from which, with appropriate **BP**s, the generalizations of psychological theories (**T_R**s) are logically derivable? In light of this reformulation, physicalism asserts that theories from physical science will occur at the end of chains of intertheoretic reductions, from psychology down to e.g., neuroscience (with intermediaries possibly in between). And each link in this chain must meet the demands that Nagel's account places on the intertheoretic reduction relation, namely, logical derivation with the help of bridge principles to connect the disparate vocabularies (in these obviously heterogeneous cases.)

What does the Nagel-supplemented IR reformulation accomplish? What do we achieve by reformulating the traditional mind-body problem in this fashion? First, we replace the murky notion of "ontological reduction" with intertheoretic reduction, a rigorous, scientifically grounded alternative. Physicalists now can appeal to examples of reduction from the history of science to illuminate and compare the nature of purported relations between theories from psychology and the physical sciences. Taking intertheoretic reduction as the central issue provides clear and justified verdicts about the variety of philosophical arguments brought to bear on the traditional mind-body problem. Evidence and arguments relevant to deciding for or against predicted future intertheoretic reductions in science are legitimate; evidence and arguments irrelevant to these issues are not.[5] This methodological lesson was taken to be significant by scientifically inspired philosophers. Philosophical questions about mind could finally be addressed from the perspective of a rigorous philosophy of science.

3 REVOLTS AGAINST NAGEL'S ACCOUNT

3.1 "Radical" Empiricists (and Patrick Suppes)

Unfortunately, just as philosophers of mind began applying Nagel's theory of intertheoretic reduction to the mind-body problem, that theory came under decisive attack within the philosophy of science. Constructed within the logical empiricist program, Nagel's theory incorporated that program's strengths and weaknesses *qua* philosophy of science. One weakness was stressed increasingly throughout the 1960s: its assumption about the continuity of scientific progress. Reduction as deduction of reduced theory **T_R** from reducing theory **T_B** reflects this assumption. Propositional logic requires that if some principles of **T_R** (the conclusion of the logical derivation) are false, then at least one of the premises must be false. Careful historical analysis—an emerging cottage industry throughout 1960's philosophy of science—of some "textbook" scientific reductions revealed that principles of $\mathbf{T_R}$s are often false. Falling bodies near the surface of the earth do not exhibit uniform vertical acceleration over any finite interval. Yet this assumed uniformity is central to Galilean physics. Galilean physics is empirically false. It does not describe correctly the behavior of falling objects in any portion of the actual world. Yet the reduction of Galilean physics to Newtonian mechanics is a "textbook" historical case. On Nagels' account, however, this requires some premise of the deduction to be false, some law or generalization of Newtonian mechanics or some bridge principle linking a term unique to Galilean physics to some Newtonian counterpart. This requirement stands in contradiction to their assumed truth (at least at the time the reduction was achieved).

As careful history of science flourished, philosophers noticed that even the case that Nagel used to illustrate his theory turned out to involve a significantly false reduced $\mathbf{T_R}$. The equilibrium thermodynamics-to-statistical mechanics and microphysics reduction is actually a limit reduction, and the limits in which the laws of equilibrium thermodynamics can be derived are never actually realized (e.g., an infinite number of gas particles whose diameter divided by the average distance between particles is only negligibly greater than zero). At best, equilibrium thermodynamical explanations approximate the actual microphysical events and their statistical distributions. Second, many key thermodynamical concepts fragment into distinct statistical mechanical/microphysical concepts. Clifford Hooker (1981) nicely demonstrates this point for 'entropy.' Third, a diachronic view of this case (its development over actual historical time) reveals mutual feedback between reduced and reducing theories. Problems confronting classical thermodynamics (the reduced theory $\mathbf{T_R}$) spurred the application and development of statistical approaches. And the "injection" of statistical mechanics (part of the reducing theory $\mathbf{T_B}$) back into classical thermodynamics resulted in more accurate predictions.[6]

What consequences do these features have for the explanations, categories, and theoretical posits of classical thermodynamics? Hooker is explicit: "In a fairly strong sense thermodynamics is simply conceptually and empirically wrong and must be replaced" (1981, 49). This quote reflects one important criticism of Nagel's account of reduction. Intertheoretic reductions in actual science typically *correct* the reduced $\mathbf{T_R}$s. Beyond a point, these corrections make untenable the "reduction as deduction of the $\mathbf{T_R}$" account.

Emerging logical and historical problems for Nagel's account spurred a number of alternative theories of intertheoretic reduction. Yet even before these problems became prominent, and spurred by some general criticisms of the underlying logical empiricist account of theory structure, Patrick Suppes (1956, chapter 12, 1965) characterized scientific theories semantically, as sets of models sharing some set-theoretic structure. In turn he characterized intertheoretic reduction not as syntactic derivability but as set-theoretic isomorphism (the formal analog of "sameness of structure"). Karl Popper (1962), Paul Feyerabend (1962) and Thomas Kuhn (1962) all developed independent accounts of reduction that insisted (or at least implied) that besides explaining why the $\mathbf{T_R}$ works in situations where it does, the $\mathbf{T_B}$ must also explain why the $\mathbf{T_R}$ fails in other expected applications. This is their sense in which the reducing theory corrects the reduced. Feyerabend (1962) famously expressed this contention by denying that reduction involves deduction in any capacity. Instead, he insisted that "incommensurability" between $\mathbf{T_R}$ and $\mathbf{T_B}$ and "ontological replacement"—a new account of the ultimate constituents of the universe—were the central intertheoretic relations. He went so far as to call for philosophers of science to abandon the search for any formal or "objective" account of intertheoretic reduction or scientific progress. Any attempt to characterize reduction or progress formally would bastardize the actual history of scientific practice.

In keeping with the trend in 1960's philosophy of mind, Suppes and Feyerabend both applied their alternative accounts of scientific reduction to the reformulated mind-body problem. Suppes writes: "The thesis that psychology may be reduced to physiology would be for many people appropriately established if one could show that for any model of a psychological theory it was possible to construct an isomorphic model within physiological theory" (1965, 59). Feyerabend writes:

> In the course of the progress of knowledge, we may have to abandon a certain point of view and the meanings connected with it—for example, if we are prepared to admit that the mental connotations of mental terms may be spurious and in need of replacement by a physical connotation according to which mental events, such as pain, states of awareness, and thoughts, are complex physical states of either the brain or the central nervous system, or perhaps the whole organism. (1962, 30)

He advocated this view for "all so-called mental states." Armed with Feyerabend's account of scientific reduction, the philosopher of mind's agenda is to "develop a materialistic theory of human beings." Such a result would "force us to abandon the mental connotations of the mental terms, and we shall have to replace them with physical connotations" (1962, 90). With

Feyerabend's account of scientific reduction, "eliminative materialism" received its first serious expression and defense.

Eliminative materialism remains deeply controversial in philosophy of mind. It has few current adherents. This current status makes it enlightening to look back at the writings of some famous "central state" or "mind-brain identity" theorists throughout the mid-1960s. As their arguments came under attack, Feyerabend's early eliminativism and the "radical empiricist" philosophy of science that underwrote it looked increasingly attractive. For example, just four years after his influential (1959) paper that brought the mind-brain identity theory to a wider philosophical audience, J.J.C. Smart claims to be

> attracted to P.K. Feyerabend's contention that in defending materialism we do not need to show its consistency with ordinary [psychological] language, any more than in defending the general theory of relativity we need to show its consistency with Newtonian theory. ... Feyerabend is perhaps therefore right in arguing that the scientific concept of pain does not need to be (and indeed should not be) even extensionally equivalent with ordinary language. (1963, 660)

Four years later Smart clarified his (cautious) change in view. He admitted to being even closer to Feyerabend, both in philosophical methodology and eliminativist conclusion. In an attempt to stave off an "ordinary language" criticism of his famous "topic-neutral translation" argument for the mind-brain identity theory, Smart writes:

> I am even doubtful now whether it is necessary to give a physicalist analysis of sensation reports. Paul Feyerabend may be right in his contention that common sense is inevitably dualistic, and that common sense introspective reports are couched in a framework of a dualistic conceptual scheme. ... In view of Bradley's criticisms of my translational form of the identity thesis, I suspect that I shall have to go over to a more Feyerabendian position. (1967, 91)

Smart was not the only famous identity theorist attracted to Feyerabend's philosophy of science and eliminative materialism. In a postscript written ten years after his famous essay defending an identity theory, Herbert Feigl writes: "I now agree with Smart (and perhaps with Feyerabend) that within the conceptual frame of theoretical natural science genuinely phenomenal (raw feel) terms have no place" (1967, 141). He cites a scientific analogy that became prevalent in later eliminativist writings: the properties of common sense physical objects vis-à-vis their "successor concepts" from microphysics. He concludes that "the phenomenal predicates used in the description of after-images, sensations, feelings, emotions, moods, etc., are *to be replaced by* the (as yet only sketchily known) neurophysiological and ultimately microphysical characterizations" (1967, 141-142; my emphasis). Thus there was a time, not too long ago, when influential physicalist philosophers of mind did not reject eliminative materialism out of hand, or the theory of reduction that supported it. It is interesting that these writings have been lost among recent philosophers of mind, who do tend to reject eliminative materialism out of hand.

These shifts by famous mind-brain identity theorists toward Feyerabend's philosophy of science (theory of scientific reduction) and eliminative materialism hold two lessons for us. First, they demonstrate how intractably dualistic our psychological concepts appear to be. It is not just extremely difficult to find a meaning preserving "translation" for mental terms within a physical language. It is probably impossible. So if we aspire to an explanation of human behavior in physical (i.e., neuroscientific) terms, we might do well to leave psychological conceptions to one side and not let them bias our views about what a successful neuroscience must look like or answer to. Second, these capitulations by identity theorists to Feyerabend show once again how attracted physicalist philosophers of mind have been to resources from the philosophy of science. Every account of intertheoretic reduction that philosophers of science took seriously was subsequently adopted by philosophers of mind in attempts to reformulate the traditional mind-body problem.

Consensus in Anglo-American philosophy of science, however, was less sympathetic to Feyerabend. Most found his accounts too radical, too dismissive of precise, formal resources to illuminate scientific concepts and historical episodes. Even so radical a critic of orthodox logical empiricism as Thomas Kuhn (1962) maintained that the majority of scientists spend the majority of their careers doing "normal science," i.e., puzzle-solving within an accepted paradigm. Parts of Kuhnian "normal science" sound like the views advocated by logical empiricists. Even cases of wholesale theory change or "scientific revolution" seem to approximate the formal intertheoretic relations proposed by Nagel and other logical empiricists. Feyerabend dismissed the possibility of accounting for this sense of "approximation," but many Anglo-American philosophers of science proceeded on the assumption that something like it could be clarified. At this point, the intertheoretic reduction literature in the philosophy of science diverged in multiple directions. I am going to trace only one of these: the search for accounts that incorporated weakened Nagelian conditions designed to capture features of scientific history and practice emphasized by Feyerabend, Kuhn and other radical empiricists. I am fully aware that other themes were stressed in the reduction literature, by luminaries like Lawrence Sklar, Thomas Nickles, William Wimsett, and Robert Causey. I leave it to fans of these philosophers to compare and contrast features of psychology's relation to neuroscience built on one of these alternatives, to the one I am about to develop.

3.2 Schaffner's General Reduction (-Replacement) Paradigm

Kenneth Schaffner's (1967) General Reduction Paradigm, later developed more fully and renamed the General Reduction-Replacement (GRR) model (Schaffner 1992), was the first explicit attempt to conciliate these features. Schaffner's model includes conditions of intertheoretic connectability and derivability (of reduced theory $\mathbf{T_R}$ from reducing theory $\mathbf{T_B}$) that yield Nagel's exact conditions as a special case. But it also includes "corrected" versions of reducing and reduced theories ($\mathbf{T_B}$* and $\mathbf{T_R}$*, respectively). This weakens the general conditions, enabling connectability and derivability to hold between $\mathbf{T_B}$* and $\mathbf{T_R}$* in cases where (actual) $\mathbf{T_B}$ corrects (actual) $\mathbf{T_R}$ by making more accurate predictions in the latter's domain of application (Schaffner 1992, 321). Furthermore, $\mathbf{T_R}$ and $\mathbf{T_R}$* stand in a relation of "strong analogy." Hence since corrected $\mathbf{T_R}$* is derivable from $\mathbf{T_B}$ (or $\mathbf{T_B}$*) (Schaffner's weakened condition of derivability), the reducing theory indicates why the reduced "worked as well as it did historically" and explains the reduced theory's domain "even when $\mathbf{T_R}$ is replaced" (Schaffner 1992, 321). These weakened notions thus "allow the 'continuum' ranging from reduction as subsumption to reduction as explanation of the experimental domain of the replaced theory" (Schaffner 1992, 320). Cases that closely approximate Nagel's conditions group around the first pole; cases with features emphasized by the radical empiricists group around the second. Both orthodox logical and radical empiricist intuitions are thereby accommodated. (However, it must be noted explicitly that Schaffner has yet to explicate the relation of "strong analogy" between corrected $\mathbf{T_R}$* and actual $\mathbf{T_R}$.)

One of Schaffner's arguments in support of his GRR model will sound familiar. He writes: "The flexibility of the GRR model is particularly useful in connection with discussions concerning current theories that may explain 'mental' phenomena" (1992, 320). The IR reformulation of the mind-body problem lives on! Schaffner shows in some detail how his model applies to an example of (potential) reduction of actual psychology to neurobiology, describing the cellular mechanisms of short-term and long-term learning as revealed by studies on the sea slug, *Aplysia californica* (1992, 323-329).[7] Although the cellular explanations Schaffner discusses are now somewhat dated (we'll take up recent discoveries in Chapter Three), the lessons he stresses for psychoneural reductions remain topical and directly relevant for my current concern. According to Schaffner (1992), the following are key features of this specific case:

- Reduced and reducing theories do not involve laws akin to those in "textbook" cases of reduction from physics (329).
- The reducing *complex* is an intricate system of causal generalizations with varying scopes of applicability (from nervous systems in general to specific types of neural processes). These generalizations are not framed within the vocabulary of one specific science (e.g., biochemistry), but rather are characteristically interlevel (containing terms from, e.g., biochemistry, molecular biology, cellular neurophysiology, neuroanatomy, and behavioral psychology) (330).[8]
- When a phenomenon described at one level (e.g., the behavioral, as "sensitization") gets explained in lower-level terms (e.g., cellular mechanisms), the former description is mapped into the lower-level vocabulary (330-331).

It is by virtue of this last feature that Schaffner's GRR conditions of connectability and derivability are generalizations of Nagel's. At first glance, Schaffner's GRR model appears supple enough to handle the special complexities and details that actual psychology-to-neurobiology reductions generate; yet it retains Nagelian-inspired conditions on the intertheoretic reduction relation. And it achieves this consilience using an example that emphasizes actual scientific details (at least as they stood in the late 1980s) far beyond the extent that is typical in philosophy of science or mind.

3.3 Hooker's General Theory of Reduction

Clifford Hooker (1981) offers another account that amounts (in part) to a weakened set of Nagel-inspired conditions. Like Nagel and Schaffner, Hooker insists that intertheoretic reduction involves deduction, with the reducing theory $\mathbf{T_B}$ serving among the premises. But unlike Nagel, the conclusion of the derivation is not the reduced theory $\mathbf{T_R}$; and unlike Schaffner, it is not a corrected version $\mathbf{T_R}^*$ of $\mathbf{T_R}$. Instead, what gets deduced is an *image* $\mathbf{I_B}$ of $\mathbf{T_R}$, specified within the conceptual framework and vocabulary of the reducing theory $\mathbf{T_B}$. $\mathbf{I_B}$ matches the domain of application and explanatory scope of $\mathbf{T_R}$ and the generalizations comprising the former mimic the logical (syntactic) structure of the latter. Simplifying, and ignoring some niceties, we can express Hooker's account with the following schematic, which I adapt with minor changes from Paul Churchland (1985):

$\mathbf{T_B}$ (& boundary conditions, limiting assumptions, as needed)
logically entails
$\mathbf{I_B}$ (a set of theorems of [restricted] $\mathbf{T_B}$)
e.g., $(x)(Ax \rightarrow Bx)$, $(x)((Bx \,\&\, Cx) \rightarrow Dx)$
which is relevantly isomorphic to ("analogous to")
$\mathbf{T_R}$
e.g., $(x)(Jx \rightarrow Kx)$, $(x)((Kx \,\&\, Lx) \rightarrow Mx)$

The schema is meant only to illustrate the "analog relation" between $\mathbf{I_B}$ and $\mathbf{T_R}$. It is not intended to provide or even ground an analysis of the relation (which, incidentally, neither Hooker nor Churchland has ever provided). "Boundary conditions" and "limiting assumptions" restrict the applicability of the $\mathbf{T_B}$'s generalizations so as to isolate falsity in the $\mathbf{T_R}$. Consider the reduction of false Galilean physics to Newtonian mechanics (presumed true at the time of the reduction). We can conjoin with the Newtonian principles either a counterfactual assumption describing conditions near the surface of the earth that permit uniform vertical accelerations over a finite interval, or a counterfactual assumption that limits the applicability of Newton's laws to moving bodies that only fall distances negligibly greater than zero. From this reducing complex—Newtonian mechanics $\mathbf{T_B}$ and the counterfactual boundary condition or limiting assumption—the image ($\mathbf{I_B}$) that mimics the explanatory scope of Galilean physics ($\mathbf{T_R}$) can be derived. The falsity of the latter is explained by and hence confined to the counterfactual boundary condition or limiting assumption.

It is important not to confuse Hooker's deduced image $\mathbf{I_B}$ with Schaffner's corrected version of the reduced theory $\mathbf{T_R}^*$. Hooker's $\mathbf{I_B}$ is characterized completely within the framework and vocabulary of $\mathbf{T_B}$; Schaffner's $\mathbf{T_R}^*$ is a corrected version of $\mathbf{T_R}$, and so is characterized (at least partly) out of the resources and vocabulary of the reduced theory. This difference yields the very different ways that Hooker and Schaffner attempt to capture radical empiricist insights within a modified Nagelian account of intertheoretic reduction. Every topic discussed in the remainder of this section contrasting the two accounts hinges on this difference.[9]

Hooker readily acknowledges the radical empiricist insights built into his account about science's actual history and progress. But his guiding intuition about intertheoretic reduction is explicitly Nagelian: "While the construction of $\mathbf{I_B}$ within $\mathbf{T_B}$ might be a complicated affair—[boundary conditions] might be fearfully complex (cf. biological reductions), counter-factual (e.g., assume continuity), necessarily counterfactual qua realization (e.g., "force free"), and so on—the ultimate relation between $\mathbf{T_B}$ and $\mathbf{I_B}$ remains straightforward deduction" (1981, 49). Even his justification of this feature is Nagelian: "$\mathbf{T_B}$ continues to directly explain $\mathbf{I_B}$ and this is the basis for $\mathbf{T_B}$'s indirect explanation of $\mathbf{T_R}$'s erstwhile scientific role" (ibid.). According to Hooker, deduction is necessary to capture the explanatory unity of $\mathbf{T_R}$ and $\mathbf{T_B}$, which remains an explicit condition of his account. This is a condition he shares with logical empiricists like Nagel.

Notice that, unlike either Nagel's or Schaffner's account, the premises of the deduction partly constituting a Hooker reduction do not contain bridge principles or correspondence rules. These are not needed. Image $\mathbf{I_B}$ is already specified within (a restricted portion of) the $\mathbf{T_B}$. There are no disparate vocabularies to bridge across premises and conclusion in the deductive component. Structures analogous in some ways to Nagelian bridge principles appear in a second stage of a Hooker reduction, involving $\mathbf{I_B}$ and the $\mathbf{T_R}$. But these components are only ordered pairs of terms that indicate the substitutions in $\mathbf{I_B}$ that yield the actual generalizations of $\mathbf{T_R}$, or approximations of the actual generalizations if that is all that a given case permits. By themselves, these ordered pairs imply neither synonymy (sameness of meaning) nor coextension (sameness of reference) of terms, nor material identity. Thus one central difficulty with Nagel's account vanishes: that of specifying the logical status of bridge principles in reductions that falsify the $\mathbf{T_R}$.

Earlier in this section we saw that Schaffner's (1992) generalizations of Nagel's conditions of connectability and derivability yielded a spectrum of possible reduction outcomes, ranging from ones in which Nagel's actual conditions are closely approximated to others displaying features emphasized by radical empiricists. Hooker's account yields a similar spectrum, ranging from "relatively smooth" to "extremely bumpy" intertheoretic reductions. A case's location on this spectrum depends on the "amount of correction" implied to the $\mathbf{T_R}$, which in turn amounts to the "closeness of fit" obtaining between the derived image $\mathbf{I_B}$ and it. Cases approximating Nagel's conditions fall near the "smooth" endpoint. (However, it remains crucial to bear in mind that on Hooker's account, the $\mathbf{T_R}$ itself is never deduced, not even in the "smoothest" cases. Rather, it is always the target of a kind of complex mimicry.) The derived $\mathbf{I_B}$ mimics $\mathbf{T_R}$'s explanatory scope in the latter's domain of application, is strongly analogous in logical structure to the $\mathbf{T_R}$, and its derivation from the reducing theory $\mathbf{T_B}$ requires few counter-to-fact boundary conditions or limiting assumptions. Historically, the physical optics-to-electromagnetism reduction seems to reflect these conditions. Cases involving features emphasized by radical empiricists fall toward the bumpy endpoint. Here an $\mathbf{I_B}$ only weakly analogous to $\mathbf{T_R}$ can be derived from $\mathbf{T_B}$, and this only with the help of numerous and wildly counterfactual boundary conditions and limiting assumptions. Historically, the phlogiston theory-to-oxidative chemistry reduction seems to reflect these conditions. "Mixed" reductions sharing some features of both extremes fall on the spectrum separating these two endpoints.

Ambiguous historical cases for Nagel's logical empiricist account, like the equilibrium thermodynamics-to-statistical mechanics and microphysics reduction, seem to reflect these conditions.

In short, a case's location on Hooker's intertheoretic reduction spectrum depends on the "amount of correction" implied to the reduced theory $\mathbf{T_R}$, captured in two conditions: the strength of analogy between the syntactic structures of the generalizations comprising $\mathbf{I_B}$ and $\mathbf{T_R}$, and the number and counterfactual nature of the boundary conditions and limiting assumptions necessary to derive such an $\mathbf{I_B}$ from $\mathbf{T_B}$.

If we are to employ Hooker's account to reformulate the philosophical mind-body problem, then it must also provide some account of when cross-theoretic identities are justified. Otherwise the ontological aspect of the traditional problem cannot be captured. Unlike Nagel's and Schaffner's accounts, Hooker's doesn't employ cross-theoretic bridge principles **BP**, the obvious resource for providing this. Instead, a reduction's relative smoothness justifies cross-theoretic identities. For not only do historical intertheoretic reductions line up on a spectrum; so do the cross-theoretic ontological consequences drawn from them. The latter range from entity and property/ event identities (visible light *is* electromagnetic radiation with wavelength between 0.35-0.75 μm) to significant conceptual revision (temperature T of a gas is only identical to its mean molecular kinetic energy in an empirically unrealizable mathematical limit) to outright elimination (there is no such thing as phlogiston). When we lay out the location of historical reductions along these two spectra—the intertheoretic reduction "amount of correction" spectrum and the ontological consequences spectrum—we find a rough isomorphism. A case's location on the intertheoretic reduction spectrum correlates closely with its location on the ontological consequences spectrum.[10]

This observation suggests a strategy for predicting the cross-theoretic consequences of a developing or potential intertheoretic reduction. First discover where on the intertheoretic reduction spectrum the case appears to be headed, in terms of the "amount of correction" being implied to $\mathbf{T_R}$. How equally explanatory and structurally analogous to $\mathbf{T_R}$ is an image $\mathbf{I_B}$ derivable within $\mathbf{T_B}$? How numerous and wildly counterfactual are the boundary conditions or limiting assumptions needed to effect the derivation? Which historical reductions does the case seem most closely to resemble in these respects? Answers will locate the developing or potential case on Hooker's intertheoretic reduction spectrum. The predicted cross-theoretic ontological conclusions (identity, revision, elimination) will then be those obtaining at the roughly isomorphic location on the other spectrum. The isomorphism across the two spectra that grounds this strategy is inspired directly by actual reductions from the history of science. Even the "autonomist" or anti-reductionist about $\mathbf{T_R}$ and its posits can be satisfied. His or her position predicts that the reduction relation won't obtain, i.e., that no appropriate $\mathbf{I_B}$ will be derived within $\mathbf{T_B}$ that is equipotent to $\mathbf{T_R}$.

Notice finally that the spaces between the "smooth" and "bumpy" endpoints on the intertheoretic reduction spectrum and the "retention" and "replacement" endpoints on the ontological consequences spectrum provide for the possibility of "revisionary" results (Bickle 1992b, 1998, chapter 6). A variety of historical scientific reductions serve as useful precedents: e.g., classical thermodynamics-to-statistical mechanics and microphysics and classical mechanics-to-general relativity theory. In both cases, the generalizations of $\mathbf{T_R}$ at best are approximated by those of the equipotent image $\mathbf{I_B}$ derivable within $\mathbf{T_B}$. Key explanatory posits of $\mathbf{T_R}$ fragment into a variety of different posits of $\mathbf{T_B}$ in different explanatory contexts. And mutual co-evolutionary interaction between $\mathbf{T_R}$ and $\mathbf{T_B}$ increases the explanatory scopes of both. Concerning the theoretical posits of psychology (the reduced theory $\mathbf{T_R}$ in the envisioned psychoneural reductions), revisionary physicalism predicts enough conceptual change to rule out cross-theoretic identities with neurophysiological posits. It differs in this way from a mind-brain identity theory. However, revisionary physicalism also denies that psychological kinds will undergo the radical elimination that befell, e.g., phlogiston and caloric fluid. Instead, one group of cognitive representation concepts (the kinds employed in psychological explanations of behavior) will be replaced by a different group *of cognitive representation concepts* (the kinds emerging from developing neuroscience). Exactly this type of result obtains in historical revisionary reductions. Relativity theory still posits length, mass, velocity concepts, just not the specific ones of classical mechanics. If revisionary psychology-to-neuroscience reductions obtain, these will yield enough conceptual change to rule out strict cross-theoretic psychoneural identities. But they will not yield wholesale elimination of psychological kinds akin to the caloric fluid/phlogiston variety.[11]

An IR reformulation of the traditional philosophical mind-body problem grounded on Hooker's theory of intertheoretic reduction looks promising. Perhaps it affords the best resources for articulating physicalism and defending it against classic philosophical objections. This remains to be seen, however, because Hooker's account (as presented so far) faces three outstanding problems. First, Hooker himself nowhere applies his account to detailed potential reductions of actual psychological to neuroscientific theories. Second, we've seen nothing in a Hooker-inspired IR reformulation (as presented so far) that addresses the most influential philosophical criticism of psychophysical reduction, the "multiple realizability" argument. Third, Hooker's theory of reduction is subject to two serious criticisms from within the philosophy of science. It is hand waving about detailed applications to historical cases of scientific reduction, leaving the key concept of an image $\mathbf{I_B}$ and the "analog relation" between $\mathbf{I_B}$ and $\mathbf{T_R}$ without a clear illustration. And as Hooker (1981) himself admits, his "analog relation" lacks precise formulation. "New wave" psychoneural reduction seeks to redress these shortcomings (Bickle 1998).

4 EXTENDING HOOKER'S INSIGHTS: NEW WAVE REDUCTION

4.1 Handling Multiple Realizability

As physicalists about mind began adopting intertheoretic reduction, Hilary Putnam and Jerry Fodor (among others) began emphasizing the problem of *multiple realizability*. Putnam published a number of papers on this topic throughout the 1960s.[12] Fodor ([1974] 1981; 1975, chapter 1) extended these arguments. They contend that a given mental type (property, state, event) is, or at least can be, realized by a variety of distinct physical kinds sharing nothing significant at the level of physical description; and, more controversially, that this fact spells doom for psychoneural reduction (and for psycho*physical* reduction generally). Putnam's favorite example was pain. The same pain state seems ascribable to creatures with very different nervous systems: humans, rats, octopi, etc. Perhaps the same pain state can be ascribed even to beings lacking earthly nervous systems, whose physical mechanisms differ completely: silicon-based space aliens, appropriately programmed digital computers, etc. But then any postulated type-identity or "reduction" of pain to a single one of its multiple physical realizers is false.[13]

The multiple realizability argument remains central to *nonreductive physicalism*, the orthodoxy in current philosophy of mind (see, e.g., LePore and Loewer 1989, Horgan 1993). If reductionism is to be a live option, the multiple realizability argument must be addressed. Can Hooker's account of intertheoretic reduction help?

Potentially, yes. Most initial replies to Putnam and Fodor appealed explicitly to resources from theories of intertheoretic reduction. Robert Richardson (1979) pointed out that Nagel (1961) himself countenanced merely conditional bridge principles. Although Nagel's detailed examples employed biconditional ("if and only if") bridge principles, all the "connectability" that his condition of derivability requires is one-way conditionals: For all x, if Bx then Rx, where 'Bx' is a predicate of the reducing theory $\mathbf{T_B}$ and 'Rx' is a predicate of the reduced theory $\mathbf{T_R}$, e.g., if x is in brain state b then x is in psychological state p. Conditional bridge principles are consistent with the multiple realizability of $\mathbf{T_R}$ posits within $\mathbf{T_B}$. Multiple realizability only nixes conditionals in the other direction: if Rx then Bx, e.g., if x is in psychological state p then x is in brain state b. So even Nagel's "classical" theory of reduction can handle the multiple realizability of reduced kinds.

Another popular reductionist reply rests upon an insight first noted by David Lewis (1969). Intertheoretic reductions are typically domain-specific. Lewis himself offered a common sense, non-scientific example to illustrate his observation, but Berent Enç (1983) and Patricia Churchland (1986), among others, have pointed out that domain specificity obtains in the "textbook" reduction of classical thermodynamics to statistical mechanics and microphysics. Temperature *in a gas* is mean molecular kinetic energy. Temperature *in a solid* is a different statistical mechanics/microphysical property, mean maximal molecular kinetic energy, because the molecules in a solid are bound up in lattice structures and restricted to a range of vibratory motions. And so on for other physical states (e.g., plasmas). Temperature is multiply realized in distinct statistical mechanical/microphysical states, and yet it is a central reduced kind in a paradigm intertheoretic reduction from the history of science. Clearly, multiple realizability *by itself* is not sufficient to block a scientific reduction—unless one is willing, on philosophical grounds alone, to part with scientific practice and deny that classical thermodynamics-to-statistical mechanics and microphysics is a "genuine" intertheoretic reduction. That move will strike many scientifically inspired philosophers as privileging prior epistemological commitments over the endeavor one seeks to understand. Jaegwon Kim (1993) even builds this domain specificity directly into his concept of reductive bridge principles. In "local reductions" the cross-theoretic bridge principles have the form, "For all x, if Sx, then Bx if and only if Rx," where 'Sx' is a predicate denoting a type of structure, e.g., if x is *homo sapiens*, then x is in brain state b if and only if x is in psychological state p. Multiple realizability only implies that different 'Bx's will occur in the embedded biconditional for different structure types, but this is consistent with structure type-specific "local reductions." According to Kim, local reductions "are the rule rather than the exception in all of science, not just in psychology" and are "reductions enough ... by any reasonable scientific standard and in their philosophical implications" (1993, 257).

However, this strategy does not handle all types of multiple realizability. Since Fodor ([1974] 1981), psychophysical anti-reductionists have emphasized a more radical sense of the relation. Call the sense introduced by Putnam "multiple realizability across physical structure types," in that distinct types of physical structures are said to realize a given mental kind differently. The domain specificity reply disarms an anti-reductionist argument employing this sense. The multiple realizability premise might be true, but the anti-reductionist conclusion does not follow validly from it. But now consider "multiple realizability within a token system across times," in which a single instance of a cognitive system realizes a given mental type in different physical states at different times.[14] The plasticity of mammalian brains—in responding to trauma, changing task demands, developmental processes, and the neural mechanisms of learning—suggests that this more radical sense is genuine. Ned Block (1978) once suggested that narrowing the scope of psychological generalizations via domain specificity to handle Putnam's sense of multiple realizability would render comparative psychology across species problematic, a seemingly legitimate endeavor. (Not to mention routine methodologies in experimental psychology and neuroscience using animal models of human capacities!). Block's point takes on added urgency when we consider that a psychology narrowed enough in its scope of applicability by domain specificity to handle the more radical sense of multiple realizability would contain generalizations applicable only to individuals at times. Surely that much domain specificity would render psychology insufficient to accommodate even the most minimal conditions on the generality of science.

However, actual scientific practice and its history give reductionists ammunition against anti-reductionist arguments built on this more radical sense. Berent Enç (1983) and I (Bickle 1992a, 1998, chapter 4) have both suggested that this radical sense is present in historical cases of intertheoretic reduction, including the paradigm classical thermodynamics-to-statistical mechanics and microphysics case. For any token aggregate of gas molecules, there is an indefinite number of microphysical realizations of a given temperature—a given *mean* molecular kinetic energy. So even this radical sense by itself appears not to block an intertheoretic reduction. Pertaining specifically to psychoneural reductions, I've argued that this radical sense is emerging in the potential reduction of propositional attitude psychology to connectionist theories of representational content (Bickle 1995a, 1998, chapter 4). These replies again grant the multiple realizability premise (interpreted in the more radical sense) and challenge the validity of the anti-reductionist conclusion said to follow from it. Their success remains conditional on accepting the standard scientific interpretation of the thermodynamics-to-statistical mechanics and microphysics case as a genuine intertheoretic reduction; but that antecedent does not seem problematic if we let scientific practice be our guide. This is the reduction cited and explained in numerous physical chemistry textbooks.

Hooker himself expresses a similar attitude toward multiple realizability in his long paper on intertheoretic reduction, writing:

> It is often argued that, e.g., cognitive psychology cannot be reduced to neurophysiology because the former cross-classifies the latter: any number of different systems (from brains to machines to leprechauns passing notes) could realize the same functional or computational theory. It helps to remove the intellectual dazzle of this fact to realize that this is true of any functionally characterized system. The same cross-classifications turn up within such prosaic fields as electrical engineering (cf. "is an amplifier of gain A" vis-à-vis particular circuit diagrams) and physics (cf. "is a high energy electron source" vis-à-vis quantum specification). (1981, 505)

His last example clearly admits of the more radical sense of multiple realizability. There are multiple distinct quantum realizations of the same token high-energy electron source over time. What is required to handle multiple realizability of psychological on physical kinds within a reductionist account is a better understanding of intertheoretic reduction across all of science. Hooker continues: "In these cases, the issue is not whether reduction is possible, but how it goes. The same applies, I hold, between theoretical domains as well" (1981, 505). Any theory of intertheoretic reduction must handle multiple realizability of reduced on reducing kinds, or that theory is insufficient for science generally, not just for psychology and neuroscience. To accommodate this feature, Hooker (1981, Part III) supplements his general theory of reduction with an account of "token-token" reduction. His account builds multiple realizability (including, potentially, in token systems across times) directly into the reduction relation. He illustrates this addition with a brief discussion of some features of the emerging reduction of transmission to molecular genetics (Hooker 1981, Part III).[15]

Each reductionist reply I've scouted accepts the truth of the multiple realizability premise and challenges the validity of the anti-reductionist conclusion. But the truth of the premise has also been challenged, especially the one employing the more radical sense. Regarding neural plasticity, regained function is often compromised following serious neural damage. Persons can still, e.g., talk, manipulate spatial representations, or move their limbs, but their performance is typically degraded. This fact gives rise to tricky questions about individuating mental types. Do these distinct pre- and post-plasticity neural events realize *the same* mental kind? This worry can be thought of as a specific version of a general challenge first raised by Nick Zangwell (1992). He claims that multiple realizability across species or structure types is "not proven" because no argument is ever given for the identity of mental types across all cases.

Kim (1993, chapter 16) suggests and I (Bickle 1998, chapter 4) elaborate that a guiding methodology in contemporary neuroscience assumes continuity of underlying physical mechanisms, both within and across individuals and species. This assumed continuity is more than mere analogy and informs most experimental techniques, research paradigms, and theoretical conclusions. Special empirical techniques and statistical analyses exist to control for idiosyncratic activity on individual trials, e.g., subtraction techniques in neuroimaging (see Posner and Raichle 1994). If the radical sense of multiple realizability really obtained to the degree stressed by anti-reductionists, the experimental techniques of contemporary neuroscience should have borne little fruit. But clearly these techniques are effective and not hopelessly naïve. Why has a detailed study of the macaque's visual system been so instructive for learning about the human's? Why has positron emission tomography (PET) and functional magnetic resonance imaging (fMRI) revealed common areas of high metabolic activity across and within individual humans, down to a current resolution of less than 1 mm and promising to go lower as techniques and analytical tools improve?[16] All of this is indirect evidence from lower level sciences that the kinds postulated by psychological theories might not be as radically multiply realized as current anti-reductionists imagine.

I now think that there is even direct evidence from recent cellular and molecular neuroscience against the truth of the multiple realiza*tion* premise that does the real work in Putnam's argument and its legacy. Neural plasticity itself turns out to be governed by common mechanisms across both individuals and species. It follows a regular progression during development, learning, and following damage to principal structures. Shared molecular mechanisms subserve it in all its forms and across the widest of gaps among biological species. I now think that this direct attack is the definitive reply to this influential anti-reductionist argument. But this attack depends upon a careful analysis of Putnam's original argument and its legacy, and upon details from the cellular and molecular investigations I will describe in the next two chapters. We thus won't be in a position to see these empirical implications for this philosophical issue until near the end of Chapter Three below. For now, you'll just have to take my word that I will attempt to cash this paragraph's promissory note.

This subsection is not intended to be a comprehensive review of the multiple realizability literature, or a definitive case for rejecting this popular anti-reductionist argument. Here I am only indicating how psychophysical reductionists have borrowed resources from both theories of intertheoretic reduction, Hooker's included, and empirical science to address this challenge. Psychoneural reduction is not Dead On Arrival. Even granted this much, however, we still confront a key problem. Is there an account of intertheoretic reduction sufficient for science generally and capable of handling the special complexities of potential psychology-to-neuroscience reductions? Or is an IR reformulation of the mind-body problem ultimately a dead-end for this less popular reason?

4.2 New Wave Reduction

So far I've stressed the attractions of Clifford Hooker's account of intertheoretic reduction. Now it is time to examine its shortcomings. Although Hooker (1981) mentions numerous historical cases when presenting his theory, he never applies it to the quantitative details of any example. Nor does he analyze or even illustrate his key concept of an image $\mathbf{I_B}$ much beyond the schema I presented in section 3 above. This omission leaves the central novelty of his account mysterious. Second, as he admits, he is unable to formulate precisely his equally central concept of the "analog relation" between $\mathbf{I_B}$ and reduced theory $\mathbf{T_R}$. This latter concept grounds the "amount of correction" implied to $\mathbf{T_R}$ via its reduction to reducing theory $\mathbf{T_B}$, and so in turn Hooker's entire intertheoretic reduction spectrum between "smooth" and "bumpy" endpoints. He laments: "Unhappily, I can think of no neat formal conditions which would intuitively separate the two" (1981, p. 223). He hints that "category-theoretic methods" might ultimately give some quantitative account of "comparative preservation indices" for a theory's "theoretically relevant properties" and subsequently for its posited entities and events (1981, p. 224). With such formal measures, one could judge quantitatively how "corrective" a given reduction is. But he admits that "all of this is very programmatic and as yet lacking in deep yet simple insight" (1981, 224). To date, Hooker has not addressed this lacuna in print.

These problems occupy crucial chapters in my first book (Bickle 1998, chapters 2 and 3). I address the first shortcoming by reformulating in Hooker-reduction terms the mathematical derivation of the classical ideal gas law (pV = nrT) from Avogadro's hypothesis, the universal gas constant, kinetic-theoretic assumptions about the nature of microparticles, and principles of Newtonian mechanics (Bickle 1998, chapter 2). One mathematical rearrangement to standard "textbook" treatments yields an equation containing only quantitative expressions from statistical mechanics and microphysics that mimics exactly the ideal gas law (Bickle 1998, 34-39). This result illustrates the structure of a Hooker image $\mathbf{I_B}$ in an actual reduction from the history of science.

To address the second shortcoming of Hooker's theory, I adopt a fundamentally different account of the structure of scientific theories (Bickle

1998, chapter 3). Hooker adopts (implicitly) the familiar "syntactic" view from logical empiricism. A theory is a set of propositions (generalizations) characterized by their syntactic structures and relations. I replace this view with a "semantic" account of theory structure, drawing on work from the "structuralist" program in philosophy of science. This program was initiated by Joseph Sneed (1971) and Wolfgang Stegmüller (1976), and was presented most completely and rigorously in Balzer *et al.* (1987). It built explicitly upon earlier work by Patrick Suppes (1956, 1965, discussed briefly in section 3 above), who characterized a theory's basic structure not as a set of sentences or propositions but rather as a set of models meeting specific set-theoretic conditions. The structuralist literature contains two developed accounts of intertheoretic reduction, one constructed self-consciously to capture Nagel's intuitions (Balzer *et al.* 1987, chapter 5), the other constructed to capture Thomas Kuhn's alternative (Mayr 1976). One explicit goal in my 1998 book was to build Hooker's concept of an image $\mathbf{I_B}$ into a structuralist-inspired account of theory structure and intertheoretic reduction. The technical (set-theoretical) details of my account are complex and don't bear repeating here, but the basic idea is straightforward. Instead of characterizing intertheoretic reduction in terms of syntactic derivations, the "new wave" approach construes the relation as the construction of an image (Hooker's $\mathbf{I_B}$) of the set-theoretic structure of models of the reduced theory $\mathbf{T_R}$ within the set comprising reducing theory $\mathbf{T_B}$.

What does this shift to semantic resources accomplish? My theory provides precise, semi-formal accounts of "amount of correction" and "location on the intertheoretic reduction spectrum." These were the very shortcomings that Hooker himself lamented. The structuralist concept of a "blur," which I extended to apply to intertheoretic relations (like reduction), provides a rough cardinal estimate of the "amount of correction" implied to the $\mathbf{T_R}$ in specific cases. Once again, I used a detailed, semi-formal analysis of the equilibrium thermodynamics-to-statistical mechanics and microphysics case to illustrate this application (Bickle 1998, chapter 3, section 5). Application of another structuralist resource, "ontological reduction links" (Moulines 1984), provided my account with an answer to a famous objection that Schaffner (1967) leveled against set-theoretic approaches to reduction. Focusing explicitly on Suppes (1956), Schaffner contended that such accounts are too weak. They cannot rule out cases of nonreduction where the theories accidentally stand in the set-theoretic relations said to comprise intertheoretic reduction. But with the "new wave" intertheoretic reduction relation characterized in part by "links" between elements of the domains and relations comprising the individual models of the reduced and reducing theories, this "too weak to be adequate" challenge is met. The "global" reduction relation across reduced and reducing theories' sets of (potential) models is itself composed of relations between the empirical base sets and relations that partially define each, relations that amount to either set identity or replacement (Bickle 1998, chapter 3, section 3). In a later chapter I reconstruct semi-formally parts of the reduction of a (schematic) propositional attitude psychological theory of cognitive representation to a (similarly schematic) connectionist-inspired counterpart (Bickle 1998, chapter 5, section 3, following Bickle 1993).

New wave psychoneural reductionism is thus the prediction that as mature theories develop in psychology ($\mathbf{T_R}$s) and neuroscience ($\mathbf{T_B}$s), images ($\mathbf{I_B}$s) of the former will be constructible within the models of latter. The challenges that Hooker (1981) was unable to meet can be addressed when his insights about scientific reduction are reconstructed and extended within an alternative account of theory structure and intertheoretic relations. This leaves only the "put up or shut up" challenge of demonstrating existing or potential reductions of actual, genuinely cognitive psychological theories to neuroscientific counterparts. I argued that the new wave reduction relation is already obtaining, albeit sketchily, between cognitivist associative learning theories and neurobiological accounts of experience-driven synaptic plasticity (Bickle 1998, chapter 5, following Bickle 1995b). This case appears to be headed for a location on the intertheoretic reduction spectrum midway between the "smooth" and "bumpy" endpoints, akin to the location of the equilibrium thermodynamics-to-statistical mechanics and microphysics case. This predicted location warrants a "revisionary physicalism" about the reduced psychological posits on the isomorphic ontological consequences spectrum.

The upshot of Bickle (1998) was that an IR reformulation of the traditional philosophical mind-body problem built upon the "new wave" account of intertheoretic reduction provides viable—and at present, the most fully developed—resources for articulating and defending psychoneural reductionism. The account is heir to a tradition running from U.T. Place and J.J.C. Smart through Ernest Nagel and Paul Feyerabend to Kenneth Schaffner, Clifford Hooker, and Paul and Patricia Churchland. In the best tradition of scientifically informed philosophy, it weaves together both philosophy of science and current empirical science, and applies the resulting mosaic to traditional dilemmas of perennial philosophy (like the mind-body problem).

5 WWSD (WHAT WOULD SOCRATES DO?)

5.1 Problems for New Wave Reductionism

Of course, not all philosophers agreed with my assessment. Ronald Endicott (2001) succinctly expresses a common criticism, writing:

> [O]ne should distinguish between *an account of scientific reduction* versus *a defense of reductive physicalism*, which, more briefly, is the difference between "reduction" and "reductionism." The latter may entail the former . . . But the former does not entail the latter, seeing that an account of scientific reduction might shed light on those cases where the world shows itself in a simple and uniform way even if, contrary to reductive physicalism, the world does not always show itself in a simple and uniform physical way. Thus I applaud Bickle's work on reduction. But I take issue with his reductionism. (2001, 378; author's emphases)

One of Endicott's arguments is especially pertinent. He argues that my adaptation of Hooker's "token reduction" relation into a structuralist framework depends entirely on an implausible eliminativism about functional kinds. Maurice Schouten and H. Looren de Jong (1999) also chide me for ignoring functionally characterized psychological concepts, especially ones employed in my favorite case from actual psychoneural science, the "memory-long term potentiation (LTP) link." They claim that this case is not an "accomplished reduction," as I claim that it is in my answer to the "put up or shut up" challenge. Instead, they insist, there remain some legitimate, not purely heuristic questions that can be answered *only* with genuinely functional concepts. Lower-level "combinatorial" explanations cannot address these (1999, 255-256). They write:

> The completed account of the molecular alphabet will only specify causal roles or dispositions, not real functions .. .When we adopt an etiological view of functions, as we have claimed one must, we cannot but conclude that a finalized specification of the dispositions of LTP is not enough to build a bridge to the functionally identified property it is supposed to explain, say, spatial learning. ... Invoking functions, to find out why a structure or mechanism exists, intrinsically refers

> to the evolutionary reasons of higher-level organization. (1999, 256)

The beat goes on. Robert Richardson (1999), drawing on themes developed by William Wimsatt (1974), insists that the rationale for pursuing research into lower level mechanisms is explanatory adequacy, not ontological parsimony. Scientists descend to lower levels only when a higher-level account fails to explain phenomena within its domain. This purported historical fact about scientific practice raises a problem for seeking ontological consequences directly from an intertheoretic reduction relation. Richardson points out further that my detailed example of a psychoneural reduction from actual current science nowhere addresses the claimed dependency or sensitivity of human-level cognition to semantic content, and so he worries that the cellular models I focus on might thereby lack sufficient resources to handle human cognition. He writes: "Without structured representations and without an understanding of the plasticity in human learning, such a reductionism will not go very far in converting the convicted cognitivist" (1999, p. 306). Barbara Hannan (2000) is even harsher on this point:

> How is a scientist supposed to capture the fact that an internal state has a representational content without using language, including sentences containing terms that refer to objects, states, and events outside the brain? Bickle's offhand reference to a neurophysiological account of content is disingenuous; we possess no such account, and he ought to know we don't (2000, 54)[17]

Finally, Achim Stephan wonders what justification I can give for why we "should look through the glass of theory reduction to come to grips with the old ontological mind-body problem" (2001, 281). Following Joseph Levine (1983) and David Chalmers (1996), both Stephan (2001) and Ansgar Beckerman (2001) see only one way to forge this connection: through an account of reductive explanation that first defines mental concepts in functional terms and then looks to intertheoretic reduction to provide the physical mechanisms that realize those functional roles. The problem with this justification is the seeming impossibility of defining some mentalistic concepts functionally (even cognitive concepts, according to Stephan). So my IR reformulation lacks appropriate justification, and according to Beckerman (2001) involves a mistaken notion of what physicalism amounts to.[18]

Notice that each of these criticisms questions, in one way or another, my application of new wave intertheoretic reduction to psychoneural examples from science. One reply we might expect from a philosopher (of science) would be to tinker with the general theory of reduction so that its application would avoid these problems. Another would be to seek out other examples from the psychological and brain sciences that better fit the general theory. (These typical responses explain the joke that serves as the title to this section.) But I'm going to take a different tack. As I've already noted, a ruthless reductionism underlies both theory and experimental practice in the cellular and molecular core of current mainstream neuroscience. So instead of massaging superficial descriptions of results from these disciplines so as to make them fit—to "apply"—a general concept of scientific reduction, I am going to delve into the details of current research and theory. Once I show how psychological theories are actually being explained in current cellular and molecular neuroscience, independent of any pre-theoretic or "philosophical" account of explanation, we can investigate the potential of this reductionism-in-practice for behavior and cognition generally. We can even examine its consequences for "hot" current topics in the philosophy of mind. As we are about to see, this move both implies and is based upon a drastic alternative to currently popular views about the nature and role of philosophy, including the view that motivated new wave reductionism.

5.2 New Wave Metascience

This book is thus an exercise in "bottom-up" philosophy of science, a project in what I will call *new wave metascience.* "Bottom up" refers to my letting a sense of reduction emerge from the detailed investigations drawn from recent scientific practice, instead of "imposing" a general account of scientific reduction onto them from science in general, or the "top down." Since in current cellular and molecular neuroscience, the devil is in the experimental and methodological details, and since these details are virtually unknown among philosophers and cognitive scientists, I will emphasize them in this book. Only then will we see *how* current neuroscience is "linking mind to molecules," and so only then can we speculate responsibly on the explanatory potential and scope of these resources for behavior and cognition generally.

One feature of my new wave metascience deserves explicit discussion. It eschews all *traditional* concern with ontology and "metaphysics," even with the secondary, derivative sense I advocated in my (1998) book. There my guiding intuition was that science is the best method we have for "doing ontology," for constructing informationally rich representations of a "real world" assumed to exist independently of our representing it. I assumed that neuroscience is doing just fine by ignoring "philosophical" objections to psychoneural reduction and proceeding instead in the fashion I tried to capture within new wave reductionism. I was urging that we accept science's strategy *tout court*. Trust scientific practice and let ontological chips fall where they may! However, I've since come to see even that project as "too metaphysical." Even though scientists talk a "realistic"-sounding language, we should not interpret this talk as addressing questions "external to" the practices and concerns of a given scientific endeavor. The job of new wave metascience is simply to illuminate concepts like reduction as these imbue actual scientific practice. To what end? *Not* to achieve some better way of addressing reformulated "external" questions about the existence and nature of "theory-independent ontology." Rather, because a reasonable explanatory goal is to understand practices "internal" to important current scientific endeavors and the scope of their potential application and development. The tasks of this book are part of a *metascience* of contemporary psychology and neurobiology, not a part of some "ontology of mind."

I borrow the terms "internal" and "external" deliberately from Rudolph Carnap ([1950], 1956). Philosophers might remember that Carnap there distinguishes two types of existence questions: "first, questions of the existence of certain entities . . . *within the* [linguistic] *framework*; we call them *internal questions*; and second, questions concerning the existence or reality *of the system of entities as a whole*, called *external questions*" ([1950] 1956, p. 206; author's emphases).[19] Using his favorite example of the "thing language," Carnap continues: "Once we have accepted the thing language with its framework for things, we can raise and answer internal questions, e.g., 'Is there a white piece of paper on my desk?,' 'Did King Arthur actually live?,' 'Are unicorns and centaurs real or merely imaginary?' and the like. These questions are to be answered by empirical investigations" (207). External questions, on the other hand, have a more "problematic character": "From these questions we must distinguish the external questions of the reality of the thing world itself, e.g., 'Does this table *really* exist?" (ibid.). According to Carnap, only philosophers ask external questions, as contrasted with "the man in the street" and scientists. The usual methods of investigation, logical analysis and empirical inquiry, do not provide complete answers to them. For these questions possess a different character than existence questions asked within a linguistic framework. He continues: "If someone decides to accept the thing language... this must not be interpreted as if it meant his acceptance of a *belief* in the reality of the thing world; there is no such belief or assertion or assumption, because it is not a theoretical question. To accept the thing world means nothing more than to accept a certain form of language" (207-208). Such a decision typically is influenced by genuinely "theoretical knowledge." What are the purposes for which the framework is to be used? How efficient is the framework for achieving these purposes? How fruitful? These questions are addressable by the usual methods of empirical inquiry. But it would be incorrect to infer that positive answers to them "support the reality of things." Instead, we should say that these features of the framework's use "made it advisable to use the physical thing language" (208)

These remarks lead Carnap to a general account of external existence questions. He writes: "Those who raise the question of the reality of the thing world itself have perhaps in mind not a theoretical question as their formulation seems to suggest, but rather a practical question, a matter of practical decision concerning the structure of our language" (207). This is not a special feature of external questions about physical things. He draws the same conclusion about external questions pertaining to numbers, propositions, thing-properties, spatio-temporal coordinate systems for physics, and the abstract objects used by semanticists in the formal analysis of linguistic meaning.

Carnap notes some profound methodological consequences of this account, especially for those seeking to undermine or "reduce" some framework to another. Consider his remarks to "nominalists" who seek to eliminate the "abstract entities" (properties, relations, propositions) commonly employed in formal semantics. Such critics must do more than simply raise problems for the "ontology" of such entities. That move only conflates internal and external questions. Instead, nominalists must "show that it is possible to construct a semantical method which avoids all references to abstract entities and achieves by simpler means essentially the same results as other methods" (221). Or consider the options Carnap makes available to skeptics of the thing language. Casting aspersions upon "the ontology of things" likewise conflates internal and external questions. Instead, skeptics must either "restrict [them]selves to a language of sense-data and other "phenomenal" entities, or construct an alternative to the customary thing language with another structure, or, finally, . . . refrain from speaking" (207). There is no short cut around *actually constructing* the alternative (linguistic) framework

I advocate a similar lesson for psychoneural reductionists. Physicalist reductionists need actually to *provide psychoneural reductions*. Paraphrasing Carnap, they must show that it is possible to construct a neuroscientific theory that avoids all references to psychological entities and achieves by simpler means the same explanatory scope as psychology. Luckily, contemporary cellular and molecular neuroscientists are doing the constructing. But these scientific developments must be set within the appropriate context. There is no shortcut for the psychoneural reductionist around this wholly empirical-cum-metascientific task. The only factual questions are "internal" ones, about the way actual neuroscientific practices are proceeding and about these practices' foreseeable explanatory potential.

Philosophers will wonder whether any kind of "internal/external" distinction can be drawn. Didn't Willard Quine trash Carnap's distinction soon after the ([1950] 1956) paper was published, by undermining any absolute distinction between the analytic and synthetic (roughly, between sentences "true by virtue of the meanings of their terms" and others "true by virtue of the way the world is")? Quine thought that he undermined Carnap's distinction. In the paper where he famously dismantled an absolute analytic/synthetic distinction, Quine writes: "Carnap has recognized that he is able to preserve a double standard for ontological questions and scientific hypotheses only by assuming an absolute distinction between the analytic and the synthetic; and I need not say again that this is a distinction which I reject" ([1949] 1953, 45-46). Two years later, Quine reiterated this point: "If there is no proper distinction between analytic and synthetic, then no basis at all remains for the contrast which Carnap urges between ontological statements and empirical statements of existence. Ontological questions then end up on a par with questions of natural science" ([1951] 1966, 134). Is any approach motivated by Carnap's distinction thereby dead in the water, in philosophy's post-Quinean times?

No. For there is at least one argument that Carnap offered in his ([1950] 1956) essay that nowhere depends upon some absolute analytic/synthetic distinction. Commenting on the classic philosophical debate between "Platonists" and "nominalists" about the existence of numbers, he writes: "I cannot think of any possible evidence that would be regarded as relevant by both philosophers, and therefore, if actually found, would decide the controversy or at least make one of the opposite theses more probable than the other" (219). From this he concludes: "Therefore I feel compelled to regard the external question as a pseudo-question, until both parties to the controversy offer a common interpretation of the question as a cognitive question; this would involve an indication of possible evidence regarded as relevant by both sides" (ibid.). Call this the "fruitless question" argument. Without any agreement about relevant evidence by disputants, a question isn't worth serious pursuit. I contend that the class of "fruitless questions" matches up closely with Carnap's original class of "external questions"—including philosophy's traditional mind-body question and even "physicalist" answers to it. Nothing in Quine's famous arguments against an absolute analytic/synthetic distinction counts against this Carnap-*inspired* distinction between (pragmatically) fruitful and (pragmatically) fruitless questions.

There is even some textual evidence that Carnap did not acknowledge the "recognition" Quine attributed to him (see again the first quote from Quine above). Throughout the entire ([1950] 1956) essay, Carnap only mentions the "analytic/synthetic" distinction once, in a discussion of distinct types of *internal questions*: those in frameworks for logic and mathematics as compared to those in frameworks for empirical science. Nowhere in his extensive discussions of external questions does he ever suggest that the analytic/synthetic distinction is involved. Furthermore, in a footnote he adds: "With respect to the basic attitude to take in choosing a language form (an "ontology," in Quine's terminology, which seems to me misleading), there appears now to be agreement between us: "the obvious council is tolerance and an experimental spirit"" (footnote 5). This hardly sounds like the "recognition" Quine claimed! Of course, this does not show that an absolute analytic/synthetic distinction isn't lurking in the background of some of Carnap's arguments. But it is very difficult to find it lurking anywhere in or around the "fruitless questions" argument I drew out of that text and expounded one paragraph above.

So I contend that we can maintain a Carnap-*inspired* distinction between pragmatically fruitful and fruitless questions even in these "Quinean" days, and that the class of pragmatically fruitless questions by and large is coextensive with Carnap's original class of "external questions." Roughly, a question is fruitful only if those advocating different answers can come to at least minimal agreement about relevant evidence toward deciding it. This distinction was not the centerpiece of Carnap's ([1950] 1956) essay, and so new wave metascience isn't merely old-fashioned logical positivism rediscovered. New wave metascience doesn't treat traditional ontological questions as "cognitively meaningless," just pointless to pursue as serious intellectual work because they amount to nothing more than disagreements over bare intuitions, with no agreed-upon evidence for resolving them. Clearly this distinction justifies my exclusive focus on practices and developments within interdisciplinary cognitive and brain sciences and my eschewing "ontological" considerations, even ones "secondary to and dependent upon" questions about intertheoretic relations, practices and attitudes within the relevant sciences. Relevant evidence is straightforward for questions "internal to" scientific practices: what experiments are being pursued, which results are being published in the best journals, which researchers (asking which questions) are getting funded, who is winning the most prestigious awards?[20] The ontological questions, however, even when reformulated using resources from philosophy of science and the relevant sciences, remain as "fruitless" as ever.

Am I being realistic? Can we really *ignore* the traditional mind-body problem? The honest must admit that the problem still exerts a pull. Can we assuage this? Again, we can learn from Carnap. Early in his corpus, he divided his "philosophical" project into "positive" and "negative" parts, writing:

> The positive result is worked out in the domain of empirical science; the various concepts of the various branches of science are clarified; their formal-logical and epistemological connections are made explicit. In the domain of *metaphysics*, including all philosophy of value and normative theory, logical analysis yields the negative result *that the alleged statements in this domain are entirely meaningless.* ([1932] 1958, 60-61)

Being jealous guardians of their historical turf, philosophers focused their critical attention almost exclusively on Carnap's negative project. Thereby they missed an important lesson buried in his positive project, and continue to do so up to the present day.

In the Introduction to his classic essay/monograph, *The Unity of Science* (1934), Carnap remarks that the "Physicalism" developed therein—that "all statements in science can be translated into physical language"—is "allied to that of *Materialism.*" But "the agreement extends only as far as the logical components of Materialism: the metaphysical components, concerned with the question of whether the essence of the world is material or spiritual, are completely excluded from our consideration" (1934, 28-29). We must reject Carnap's tools for carrying out his positive project: his quaint accounts of "translation" and "logical analysis" and his willingness to massage actual scientific practice to generate "rational reconstructions." The replacements I am urging indicate another way that new wave metascience is more than just warmed-over Carnap, despite shared anti-metaphysical motivations.[21] But I now agree wholeheartedly with the intuition guiding this quote. The "physicalism" of new wave metascience is not an "ontological hypothesis." It is instead a description of reductive practices and accomplished results "internal to" contemporary cellular and molecular neuroscience. It is an argument for the explanatory potential of these practices. It is an application, mostly debunking, of this "reductionism-in -practice" to some trendy assertions in current philosophy of mind. And it is an account of the entire multi-leveled discipline of neuroscience from the perspective of its current cellular and molecular core.

There is another sense in which new wave metascience implies changes to methodology in philosophy. Even philosophers sympathetic to science and its potential impact on philosophical issues tend to reserve a task for philosophy that is independent in at least some sense from scientific details. The status of "reduction" seems clearly to have something to do with "explanation." The reducing theory must at least explain all the data that the reduced theory explains. So doesn't any defense of reductionism require an account of explanation? And isn't that a task for philosophy—sensitive to scientific practices and results, but "over and above" them? If one isn't armed with such an account, isn't the choice of investigating cellular and molecular neuroscience, as opposed to cognitive neuroscience or even cognitive science in general, arbitrary?

New wave metascience says no. Scientists tend to do just fine with a rough-and-ready understanding of what counts as an "explanation" and what distinguishes a "good" one from a "poor" one. At the very least, the practices accepted by practitioners of successful scientific disciplines are enough to separate successful from unsuccessful programs and to self-correct errors and mistakes. (It is commonplace to hear philosophers of science remark on science's "self-corrective" nature. One guesses that scientists didn't acquire this capacity by reading philosophers of science.) Starting a philosophical analysis with a detailed example of a scientific explanation is part of "taking science at face value." There is nothing wrong with trying to make these practices more explicit, to abstract general principles out of specific instances. That is a philosophical project. But the guiding intuition of new wave metascience is that this project should be pursued from the bottom up, that is, from a detailed investigation of real examples of current research and an attempt to extend these results and practices to issues traditionally reserved for (and by) philosophers. However, there is no "magic argument" for this feature of new wave metascience. The only procedure is to start constructing from current scientific examples and then to challenge philosophers to articulate what seems to be missing; and, if an answer to the first question is offered, to ask why those missing features are important. The rest of this book is a first step toward carrying out this procedure with regard to the "ruthless reductionism" endemic in current mainstream (cellular and molecular) neuroscience. Judge this methodological feature of new wave metascience, that of trusting what scientists consider successful explanations, only after the construction has been presented.[22]

Some philosophers might worry that new wave metascience, especially its Carnap-inspired "pragmatism," concedes too much to "anti-realists" and "mysterians." Can't one mobilize pragmatic considerations directly to defend a Realistic interpretation about the results of our successful sciences? And what about philosophy itself? What is the content of my purported "positive" project? Another important philosopher from the mid-20th century, Hans Reichenbach—whose writings are now as equally neglected as Carnap's—had some pithy remarks that address both of these questions. In his more popular book, *The Rise of Scientific Philosophy* (1957), Reichenbach notes the misplaced confidence that some philosophers place on science, especially philosophers of science, as contrasted with the scientist who from his own experience knows better:

> It is a strange matter of fact that those who watch and admire scientific research from the outside frequently have more confidence in its results than the men who cooperate in its progress. The scientist knows about the difficulties which he had to eliminate before he could establish his theories. He is aware of the good luck which helped him discover theories that fit the given observations and which make later observations fit his theories. He realizes that discrepancies and new difficulties may arise at any moment, and he will never claim to have found the ultimate truth. Like the disciple who is more fanatical than the prophet, the philosopher of science is in danger of investing more confidence in scientific results than is warranted by their origin in observation and generalization. (1957, 43).

The "scientific realists" that came to dominate late-20th century philosophy of science stand guilty of Reichenbach's charge. I suppose that if one is going to do metaphysics, then scientific realism beats the alternatives. But there is the option of eschewing metaphysics for investigations internal to particular practices, where the latter are chosen to be worthy of investigation on pragmatic grounds.

This still leaves the nagging question about the content of a "positive" philosophical project aimed at mainstream (cellular and molecular) neuroscience. I've already said that providing the first step toward this is the project of this entire book, so small wonder that I can't articulate a convincing short answer to it in a few sentences at the outset. But a remark from Reichenbach is instructive about where the materials for this positive project are to be found. Commenting on the origins of the "new scientific philosophy" in 19th century science, Reichenbach writes:

> Just as the new philosophy originated as a by-product of scientific research, the men who made it were hardly philosophers in the professional sense. They were mathematicians, physicists, biologists, or psychologists. Their philosophy resulted from the attempt to find solutions to problems encountered in scientific research, problems which defied the technical means thus far employed and called for a re-examination of the foundations and the goals of knowledge. Rarely was such philosophy detailed or explicit, nor was it extended beyond the confines of the fields of their maker's particular interests. Instead, the philosophy of these

> men lurks in the prefaces and introductions of their books and in occasional remarks inserted in otherwise purely technical expositions. (1957, 123)

Regarding Reichenbach's last point, recall the quotes from the introductory chapters of the last two editions of Kandel, Jessell, and Schwartz's text that I cited early in this chapter to illustrate the "ruthless reductionism" endemic in current mainstream neuroscience. The job of Reichenbach's "scientific philosopher" is to "collect these results" of scientific research, add to them with scientific investigations (if motivated to do so), and "present them with all their interconnections." The result is a "synopsis of scientific answers to philosophical questions" (1957, 119), or at least to the residue of the questions that are fruitful to address. Scientific philosophy as it will be practiced in this book—new wave metascience—is heir to a brilliant tradition spanning most of the 19th and 20th centuries, until the end of the latter, when "speculative metaphysics" infected even philosophy of science and philosophy of particular sciences with "external," "pragmatically fruitless" debates that turn on nothing more than clashing intuitions.

My strong statements about an alternative to philosophy of neuroscience as currently practiced, and more generally to the broader current philosophical project of "naturalizing the mind," coupled with references to half-decade-old (and more) writings by out-of-vogue positivists, won't convince philosophy's orthodoxy. That isn't my purpose in this chapter. I'm here only staking out a position, making explicit the bigger picture that motivates my interest in state-of-the-art cellular and molecular neuroscience, the mainstream of the current discipline. The only arguments that should convince anyone are the ones mobilizing resources from these sciences and showing how much they can accomplish. Even if you don't share my philosophical motivations or program, or if you are neutral on these issues about philosophical method (as I suspect most neuroscientist readers are), you might still find it interesting to see how much can be said in response to some influential philosophical issues by building up from the "ruthlessly reductive" basis provided by current mainstream neuroscience. If you count yourself as a philosopher of neuroscience, you at least owe a look at the work that constitutes the bulk of your discipline's current research. And if you are an anti-reductionist about mind for any reason, how can you know the limits of "ruthless reductionism" if you don't know what that approach is actually doing and achieving in current science?

I close this introductory chapter with one more nod to a forerunner. Obviously, a work like this one owes much to Patricia Churchland's groundbreaking *Neurophilosophy* (1986). Before that book carved out a niche, no one would have thought to combine philosophy and neuroscience in the way I will here. But I think I can do Churchland one better, and not just by updating her neuroscience by a decade-and-one-half. In the General Introduction to her book, she writes:

> Philosophers who are expecting to find in the introduction to neuroscience a point-by-point guide of just what facts in neuroscience are relevant to just what traditional philosophical problems will be disappointed. I have made some occasional efforts in that direction, but in the main my eye is on the overarching question of the nature of a unified, integrating theory of how, at all its levels of description, the brain works. (1986, 7)

I too am interested in that overarching question (as a piece of new wave metascience), but I also think I can give a "point-by-point guide" of which results from recent cellular and molecular neuroscience are relevant to which current philosophical problems. Details and applications over the next three chapters will confirm or disconfirm this.

NOTES

[1] My discussion in the next two sections draws on Bickle (2000).

[2] Carl Craver has pointed out to me that stressing this observation might be useful in breaking down some of the communication barriers that still exist between neuroscientists and philosophers. Current philosophers as a group are not a horde of substance dualists—as the introductory chapters in many neuroscience textbooks suggest or imply!

[3] See Churchland, P.M. (1987, 58-59) for some simple examples.

[4] In my (1998) I took folk psychology to be the theory with which our common sense ontology of mind is affiliated. A number of commentators took me to task for this, including Bontley (2000), Hannan (2000), Richardson (1999), and Stephan (2001). I accept their criticisms (although in Bickle 2001 I try to explain why I took this view). Nothing hinges on this change for the advantages of the "Intertheoretic Reduction reformulation" of the mind-body problem about to be stressed.

[5] I treat this point in detail in my first book, including providing examples of "linguistic" arguments still in vogue in some philosophical circles today that are irrelevant to theory reduction issues in science. See Bickle (1998, Chapter 2, section 3, especially pages 44-45).

[6] Hooker (1981) provides a nice introduction to these details. I capture some of these details within a quasi-formal account of the intertheoretic reduction relation (Bickle 1998, chapters 2 and 3).

[7] This is the same example I employed, independently, in Bickle (1995b) and (1998, chapter 5). My use of this case to illustrate an actual psychoneural reduction goes back to my doctoral dissertation (Bickle 1989).

[8] With regard to both of these first two features, Schaffner is implying that not only is Nagel's account of intertheoretic reduction incapable of handling psychology-to-neuroscience theory relations, but that the very account of "theory" he presupposed is wrong for these cases, also. Thanks to John Symons for calling this point to my attention.

[9] For this reason, I've changed the symbol Hooker (1981) uses to denote the "analog structure." Unfortunately, he chooses the symbol $\mathbf{T_R}$*.

[10] In Bickle (1998, 30), I provide a diagram of these isomorphic spectra with these historical reductions located on both.

[11] In Bickle (1992b) and (1998, chapters 5 and 6) I offer empirical evidence for a future revisionary reduction of psychological to neuroscientific theories. There I argue that while a synaptic weight-vector based account of cognitive content drawn from cognitive and computational neuroscience eschews the propositional contents of belief-desire psychology, it nevertheless preserves the coarse-grained functional (cause-and-effect) profile and the intentionality that the latter ascribes to cognitive states, especially to states near the sensory and behavioral (motor) peripheries.

[12] Key papers are reprinted in Putnam (1975a). In Bickle (1999) I review key themes in the literature that Putnam initiated, but see my (1998, chapter 4) for a more detailed, technical discussion.

[13] I'll describe this argument more completely in Chapter Three, section 4.

[14] I introduced these terms in Bickle (1992a) and expounded them further in Bickle (1998, chapter 4).

[15] I explicate this account further in Bickle (1992a) and apply it to a psychoneural example in Bickle (1998, chapter 4).

[16] Trent Jerde points out to me that a great majority of PET and fMRI studies use voxel sizes in the 3 X 3 X5 mm range; voxel sizes of 1 mm or less almost always refer to anatomical images. And even this much grosser "standard working resolution" for functional images requires statistical techniques that "smooth the data" or require the presence of "contiguous active voxels" to assert a claim about significant activation. I don't disagree. There is much guff about functional neuroimaging in both the popular press and philosophical discussions. The study I'll report in detail in Chapter Three below uses both the much less impressive voxel sizes he cites and a variety of statistical "smoothing" techniques. But Scott Holland also informs me that recent high-field magnets (e.g., 7 Tesla), now approved only for research on laboratory animals, can get far smaller *functional* resolution than even 1 mm.

[17] "Offhand reference"? I spent twenty pages (in my 1998, chapter 5) explaining the psychological and neuroscientific details of associative learning, hierarchically structured memory, and experience-driven synaptic plasticity, as these theories stood in the early 1990s. I then spent most all of chapter 6 drawing consequences from these details for the realism-eliminativism debate about propositionally structured content.

[18] These are not the only challenges to new wave reductionism in print. Others will arise later in this book.

[19] Unless otherwise noted, quotes from Carnap in the remainder of this chapter are from his ([1950] 1956). I indicate them only by page number.

[20] Notice that many of my questions go beyond what Carnap considered "internal questions." Many will notice Kuhnian themes in my list. This is another sense in which new wave metascience isn't merely logical positivism revived. The distinction I am stressing is Carnap-*inspired*, not Carnap's.

[21] The philosophy of science (minus the ontology) in Bickle (1998, chapters 2 and 3) can be re-interpreted in light of new wave metascience as providing alternatives to Carnap's quaint notions and translational strategy.

[22] Thanks to Carl Craver and Dingmar van Eck for insisting that I elaborate on this important methodological difference between new wave metascience and more mainstream philosophy of science.

CHAPTER TWO
REDUCTION-IN-PRACTICE IN CURRENT MAINSTREAM NEUROSCIENCE

Most philosophers of mind and many cognitive psychologists still doubt that "genuinely cognitive" psychological theories will reduce to neurobiological counterparts. As I emphasized in Chapter One, this attitude contrasts starkly with the reductive aspirations of "mainstream" cellular and molecular neuroscientists. My first substantive task in this book is to characterize "reduction-in-practice," as it is generating experiments, results, and explanations in current mainstream neuroscience. For reasons sketched in the previous chapter, I'll ignore at first the concerns typically assumed by philosophers to occupy center stage in discussions of psychoneural reduction, namely philosophical anti-reductionist arguments and the problems of characterizing a concept of intertheoretic reduction for science generally. Instead, I'll begin by presenting a detailed example, recent discoveries about the molecular mechanisms of long-term potentiation (LTP), an important type of experience-driven synaptic plasticity, and the behavioral data these mechanisms explain. This story is an accomplished neurobiological reduction of psychology's "memory consolidation switch" that mediates the conversion of short-term to long-term memories. These recent scientific details show the nature of reduction at work and succeeding in current cellular and molecular neuroscience. Later in this chapter, as a piece of new wave metascience, I'll generalize from this example, hopefully to provide a template for additional psychoneural reductions.

There is a second reason why the details of this example are important for philosophers and cognitive scientists. There is a widespread misconception about the content of "neuroscientific explanations" of behavior. Many still think of these explanations on the model of what neuroscience provided three decades ago, namely, circuit diagrams of anatomical connections between neural regions known to subserve specific functions, with some molecules (e.g., neurotransmitters) indicated in a few spots. Indeed, there is a widespread misconception that "we don't know much about how the brain works." This misconception ignores completely the "molecular-biological wave" that began sweeping through neuroscience more than two decades ago. I gave sociological data in support of this trend in Chapter One and there suggested that more than sociological justifications for it could be offered. Now I want to make good on the latter assertion by presenting an example of a detailed neuroscientific investigation, showing how this approach has begun generating novel *behavioral* experiments, results, and explanations. This example is intended to demonstrate clearly the character of mainstream neuroscientific investigations in the present and the foreseeable future. Finally, it will show that we do know a lot about "how the brain works," at least its fundamental components, and how we can manipulate these components to produce specific, measurable behavioral effects. Even if philosophers, cognitive scientists, and cognitive neuroscientists find reasons to reject the "ruthless reductionism" I urge from this current example, they will at least see some details about actual practices in mainstream neuroscience circa the beginning of the 21st century.

1 A PROPOSED "PSYCHONEURAL LINK"

Much has been written over the past twenty years about a "learning and memory—long term potentiation (LTP) link" And much of this scientific and theoretical literature has been misunderstood. Not even its most strident proponents think that LTP is *the* cellular/molecular mechanism for all "forms" or "types" of memory (although their writings sometimes give this mistaken impression.) About the "simple proposition" that "LTP is a substrate of memory," neurobiologist Gary Lynch has written recently:

> Substituting the definite article in this proposition converts it to a form that is manifestly *not* true (it is easily shown that forms of memory exist that do not involve LTP), and yet it is widely employed, perhaps unwittingly, in LTP/behavior arguments. These difficulties provide ample reason to be pessimistic about the still-popular search for an *experimentum cruces* that "proves" or "disproves" the role of LTP in memory. (2000, 139).

Lynch has been among the most forceful and influential proponents of a "learning and memory-LTP link" for more than two decades.

Memory-like features of LTP have been explored vigorously since its discovery in the early 1970s. These include its

- *selectivity*: It only affects transmission efficacy at specific synapses on a given neuron.
- *cooperativity*: Larger effects and more stable induction occur when more afferent fibers (input lines to the neuron) are stimulated.

- *multiple forms*: Different molecular mechanisms underlie its variable stability over time, reminiscent of the short-term/long-term memory distinction.
- *cumulative nature*: Successive episodes of high-frequency stimulation in the same afferent fibers produce increasing amounts of synaptic potentiation.
- *regional distribution*: LTP has been documented in synapses in mammalian cortex, hippocampus, other "limbic" structures, and spinal cord (there called "wind up"). All of these areas have been associated behaviorally with important types of long-term memory.[1]

Attempts to correlate the temporal dimensions of these synaptic features with those from behavioral studies of memory have been prominent since LTP's discovery. The advent and dominance of molecular techniques and investigations in mainstream neuroscience over the past two decades have influenced LTP research strongly. Work on the molecular mechanisms of LTP and attempts to relate these to measurable memory-guided behavior are at present among the hottest and best-funded research areas in all of science. Some decry this fact, but no one denies it. Since I want a detailed example that reflects accurately the aims, goals, and aspirations of current cellular and molecular neuroscience, this research program is a clear choice.

I should confess at the outset to finding this program and its mechanisms to be among the most beautiful research in contemporary science. I am inspired by it and what I see as its explanatory veracity and potential. Accuracy and aesthetics aside, however, my purpose in describing it is to present a paradigmatic example of an accomplished psychoneural reduction, and hence a template for additional reductions. Like it or not, this example illustrates the force of reductionism—including its ruthlessness—at work in current cellular/molecular neuroscience.

However, I am discussing a limited proposal: that memory *consolidation* has been reduced ("linked") to the molecular mechanisms of LTP. This is the extent to which a reduction is on offer, at least in the serious scientific literature, and we'll even see in this chapter and the next a number of distinctions that have been drawn between types of memory systems. Scientists virtually never assert a global "memory is LTP" hypothesis in primary research publications. Incidentally, the limited scope of this proposal reveals errors in one of Schouten and de Jong's (1999) criticisms of my previous appeal to this example as an accomplished new wave psychoneural reduction. They write: "The functional characterization of associative, Hebbian learning in terms of weight changes of synaptic connections has driven LTP research by staging certain features of learning and memory. The molecular account would have to provide the necessary and sufficient conditions for these features if a reductive explanation was to be achieved" (1999, 249). They next point out that LTP is found in a variety of brain systems, including some that apparently have nothing to do with learning and memory. They cite approvingly the list given in McEachern and Shaw's review (1996, section 9.1) and the lesson urged there: "LTP ... serves functions other than, or in addition to, memory" (1996, p. 80, quoted in Schouten and de Jong 1999, p. 250). So the "sufficiency" condition fails. They also cite Saucier and Cain's (1995) study, where a potent and specific antagonist (blocker) of NMDA (N-methyl-D-aspartate) receptors, NPC17742 (2R,4R,5S-2-amino-4,5-(1,2-cyclo hexyl)-7-phosphonoheptano acid) completely blocked LTP induction in hippocampal dentate gyrus neurons. Nevertheless, rats treated with this receptor antagonist that had been made familiar with water maze task demands by non-spatial pre-training displayed normal spatial learning in the water maze as compared to pharmacologically untreated controls (Saucier and Cain 1995, Figures 2 and 3).[2] So LTP isn't necessary even for spatial learning and memory. Schouten and de Jong conclude, contra to my answer to the "put up or shut up challenge," that a memory-to-LTP reduction isn't yet accomplished, as these concepts now stand in psychology and neuroscience.

Two mistakes infect their criticism. First, the serious scientific hypothesis links only one process of memory—consolidation—to LTP. Second, no "necessity and sufficiency" requirements attach to reduction projects in cellular and molecular neuroscience. No serious LTP researcher is unaware of the results Schouten and de Jong cite. In fact, we'll see in the next chapter that the molecular mechanisms of LTP are not even unique to neurons, much less to neurons exclusively involved in memory consolidation! And yet the "memory consolidation-molecular mechanisms of LTP link" continues to thrive. Clearly, some other sense of reduction is at work than the one Schouten and de Jong (1999) find lurking in some neuroscientists' writings. It is this other sense that I want to bring to the surface by looking in some detail at the actual neuroscientific accomplishments and assertions, as they have been presented in the primary experimental literature (as opposed to review papers, critical surveys, and the like).[3]

2 TWO PSYCHOLOGICAL FEATURES OF MEMORY CONSOLIDATION

Consider first some feature of memory consolidation revealed by experimental psychology. These features are the explanatory targets of the reductionistic search for molecular mechanisms. Since the seminal research of Ebbinghaus, Müller, and Pilzecker more than a century ago, and then elaborated by William James in his classic *Principles of Psychology* (1890—though James employed his own terminology), psychologists have distinguished *short-term* from *long-term memory*. Short-term memory is transient, lasting anywhere from the immediate present ("iconic memory") up to several minutes ("working memory") to perhaps an hour or more (with rehearsal), and is very susceptible to interference by distraction. Long-term memory is stable, lasting for weeks, months, years, sometimes even decades, resistant to distraction, and (typically) is induced only with stimulus repetition (which can include extended rehearsal).[4]

Two features of the "conversion" of short-term memory into long-term memory have been prominent in experimental studies for more than one century. Ebbinghaus first showed that

- stimulus repetition is necessary (in most instances) to induce stable, stronger, longer-lasting memories.

He presented subjects with novel letter strings, composed of "syllables" of one consonant followed by one vowel followed by another consonant.[5] To minimize existing "associations" that subjects might have with particular strings, which could artifactually influence their retention and recall, he constructed his sequences randomly and discarded strings that formed actual (German) words. He thus constructed a novel "language" out of over 2300 syllables, from which for a given experimental run he chose 7 to 36 to construct lists of varying lengths. Subjects heard members of the lists spoken at a rate of 150 per minute during training and then had to recall the syllables that occurred on the lists at various times after training. Ebbinghaus varied the number of times a given list was repeated during a training session, from anywhere between 8 and 64 times. He discovered a nearly linear relationship between the number of repetitions of the list during learning and retention tested on the following day.

Second, as Müller and Pilzecker also first demonstrated experimentally more than one century ago,

- the conversion of information from short-term memory to long-term memory can be disrupted by *retrograde interference*, by distractions introduced *after* the initial items have been learned and stored in short-term memory.

They coined the phrase *consolidation period* to refer to the time needed for the short-term "memory trace" to achieve stable long-term form. Experimental psychologists ignored this initial work on retrograde interference for more than a half-century. Clinicians, on the other hand, were quite familiar with this effect, knowing that a head injury often erased recall of events that *preceded* the blow. Retrograde interference was then the subject of two very influential studies in the late 1940s. Consider one of these, still cited in the consolidation literature today.

Duncan (1949) had noticed during a prior study of electro-convulsive cerebral shock on "new and old learning" that the time the shock is delivered relative to the time of training had a significant influence on its effect. To study this "retroactive inhibition," he used a conditioned aversion procedure that rats learn easily. Rats were introduced into a two-compartment box with an open doorway between the compartments and initially allowed to explore both compartments for two minutes. The "test" compartment had an electric floor grid. The "safe" compartment did not have a floor grid, but was brightly illuminated. Because of their light-phobic nature, rats quickly retreat back into the test compartment during this initial two-minute exploration. At the end of the two minutes, the electric floor grid was activated, delivering a mildly aversive foot shock. Rats quickly ran into the safe compartment for the ten seconds that the grid was charged. They were removed from the box immediately.

Rats were then divided into one of eight experimental groups and a control group. Members of each experimental group received a cerebral electroshock, delivered through ear electrodes, that was sufficient to induce grand mal convulsions on each application. Each experimental group received the cerebral shock at a specified time after being removed from the training box: 20 seconds, 40 seconds, 60 seconds, 4 minutes, 15 minutes, 1 hour, 4 hours, or 14 hours. (These specific times were chosen to make for easy graphical representation of the data on a logarithmic scale.) Control animals underwent the training but received no subsequent cerebral shocks. After this initial training trial, each rat was returned to the test compartment once a day for 17 days. Ten seconds after the rat was placed in the test compartment, the electric grid was charged for ten seconds and rats received the foot shock if they had not already run into the safe compartment. The rat was then removed from the box and received the appropriate cerebral shock, depending on which group it belonged. On the eighteenth day, no subsequent cerebral shocks followed the test trial. Rats in the various groups were measured on the average number of "anticipatory runs" they made into the safe compartment during the ten seconds between being placed in the test compartment and grid charging. Significance was measured by comparison to the control averages.[6]

Duncan's results spoke clearly in favor of a "consolidation" theory of the effects of this retrograde interference, namely that "newly learned material undergoes a period of consolidation or preservation. Early in this period a cerebral electroshock may practically wipe out the effect of learning. The material rapidly becomes more resistant to such disruption" (1949, 44). Compared to controls, only animals in the 20 second, 40 second, 60 second, 4 minute, and 15 minute groups showed statistically significant reduction in the average number of anticipatory runs into the safe compartment prior to the shock. (See Figure 2.1.) The 1 hour, 4 hour, and 14 hour groups were no different than the control group even when averages were broken down into three-day segments. Furthermore, Duncan found a steep and progressive relation between the time after training that the cerebral shock (retrograde interference) was delivered and the effect on learning and memory. The effect lessened rapidly and progressively as more time elapsed.

A control experiment ruled out the possibility that this retrograde interference was simply the result of the cerebral shock being a second aversive CS-US pairing. By 1949, experimental psychologists knew that training an animal on a second conditioning task soon after training on a first inhibits learning of the first association. This possibility confounded the interpretation that the retrograde cerebral electroshock interfered specifically with neural consolidation mechanisms. Duncan (1949) trained rats on the same conditioned aversion task described above, but instead of experimental animals receiving cerebral electroshocks at some time subsequent to each training episode, they received retrograde hind leg shocks (of the same intensity as the cerebral electroshocks in the original study). If the original effect was simply one of inhibition based on a second CS-US pairing, then experimental groups in the control study should demonstrate the same pattern of behavioral inhibition as did the analogous groups in the original study. They didn't. Only the 20-second group in the control study showed any statistically significant inhibition. Every other experimental group of fifteen minutes or less performed similarly to controls, who received no retrograde leg shocks. And even the 20-second retrograde leg shock group was significantly less inhibited in learning and remembering the conditioned aversion task than their counterpart retrograde cerebral shock group.

Evidence from this control experiment supported the claim that the retrograde cerebral shocks were inhibiting learning and memory by acting directly on neural consolidation processes. Duncan concluded:

> A consolidation period follows the completion of each trial in the avoidance box. This period is less than 1 hr. and very probably is not significantly longer than 15 min. The disruptive effect of an electro-convulsive shock, when applied during the consolidation interval, produces a loss of retention, which appears as slower learning. When the shock is interpolated sooner after each trial, less time is allowed for consolidation and there is greater interference with retention.

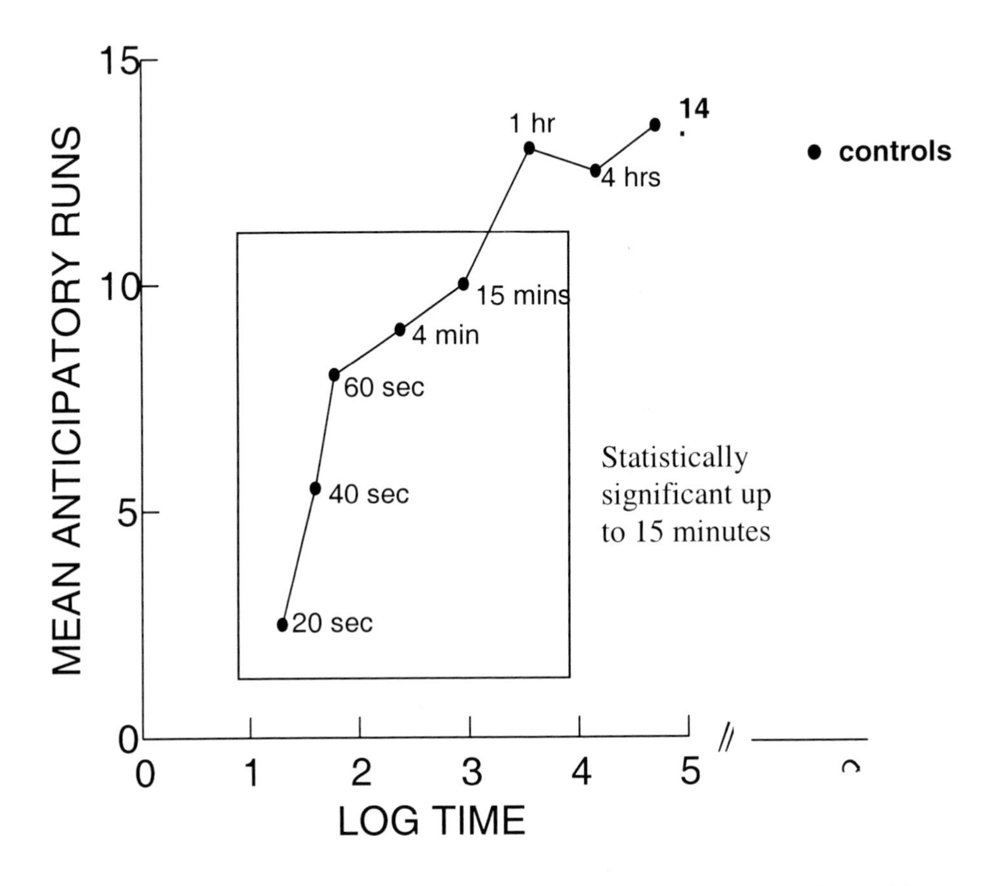

Figure 2.1. Average number of anticipatory runs into the "safe" box across all eighteen trials as a function of the trial-electro-convulsive shock time interval expressed in logarithmic units. Each point on the curve represents a different experimental group (indicated by the amount of time separating the trial and electro-convulsive shock for each trial for each member of that group). Statistically significant differences compared to control were found up through the 15-minute interval group. Graph created by Marica Bernstein. (See Duncan 1949, Figure 1.)

> When the shock is applied after consolidation has ceased, no retention loss is found. (1949, 43)

Although Duncan's study occurred during the zenith of methodological behaviorism in experimental psychology, his results fit naturally with a more recent "cognitivist" interpretation. Notice that he himself speaks of "retention, which *appears* as slower learning" in the above passage, which suggests a cognitivist interpretation. This point is important because a reduction of Duncan's results to underlying molecular mechanisms would thus be a reduction of a "genuinely cognitive" psychological process, according to the currently acknowledged "mark" of this distinction. The two key features of consolidation—the need for repetition/rehearsal and its susceptibility to retrograde interference—are naturally thought of in terms of memory ("mental") *representations* and operations involving their contents. Rehearsal amounts to "being presented" or "calling to mind" repeatedly the representational content of the stimulus presentation now available in short-term memory. Retrograde interference disrupts the needed maintenance of the memory representation. Such accounts fit with the widely accepted "mark of the cognitive," as requiring explanations that advert to operations over the contents of cognitive representations. "Autonomists" about the cognitive-psychological from the neural should thus urge that explaining the memory consolidation switch is a job for experimental psychology, not for neuroscience—and certainly not for the branch of neuroscience that deals with the biochemistry of receptor proteins, intracellular second messenger signaling pathways, neuronal gene expression, and the like.[7]

Unfortunately (for psychological autonomists, at least) subsequent psychological research on memory consolidation quickly hit explanatory limits. A few features of consolidation proved addressable at the psychological level: its variable time courses for different memory items; the nature and time courses of effective versus ineffective retrograde interference; and the variable amount of repetition/rehearsal required to consolidate different items. Most glaringly, experimental psychologists were never able to explain mechanistically the consolidation process or switch. Neuroscientists, on the other hand, quickly made impressive progress. Specific pharmacological manipulations dating back nearly forty years have produced animals (mammals included) with intact learning and short-term memory capacities but profoundly deficient long-term memory. Such manipulations seem to disrupt selectively the consolidation switch. Davis and Squire's (1984) influential review paper has now been in print nearly twenty years, yet it still describes how fruitful neuropharmacological manipulations were in elucidating the initial mechanisms of the consolidation switch. Over the past decade, in keeping with biotechnology's impact, experimenters have carried out these manipulations using genetic knockout and transgenic rats and mice. These technologies resolve nagging methodological worries that plague pharmacological experiments and the proper interpretation of their results.[8] The outcome of these neuroscientific investigations are current models of the molecular mechanisms of LTP and the measurable behavioral effects on learning, short-term memory, and long-term memory by manipulating them directly.

3 LTP IS DISCOVERED

3.1 From Hebb's neuropsychological speculations, 1949, to Norway, 1973

The current approach to learning and memory in mainstream neuroscience follows a lead first developed explicitly by psychologist Donald Hebb. (I say "explicitly" because the great Spanish neuroanatomist, Santiago Ramon y Cajal, suggested this lead more than one-half century earlier in an address to the British Royal Society.[9]) In his classic book, *The Organization of Behavior* (1949), Hebb recommended that we think of learning and memory in terms of *synaptic strength* and *plasticity*: the changeable effect that a given neuron exerts on electrochemical membrane potentials in neurons with which it shares an active synapse. He begins with a "bald assumption about the structural changes that makes lasting memory possible ... in brief, .. .that a growth process accompanying synaptic activity makes the synapse more readily traversed. ... An intimate relationship is postulated between reverberatory activity and structural changes at the synapse" (1949, 60). On purely theoretical grounds, Hebb postulated a "dual trace" synaptic mechanism for memory: first a "transient, unstable reverberatory trace," followed by a time-dependent, reinforcing "more permanent structural change" (1949, 62). As we will see, Hebb's postulation was remarkably prescient. After briefly discussing the then-current neurobiological concept of "synaptic knob growth," Hebb insists that "the details of these histological speculations are not important except to show ... that the mechanism of learning discussed in this chapter is not wholly out of touch with what is known about the neural cell" (1949, 65). But in his final analysis, Hebb was honest: "As neurophysiology, this and the preceding chapter go beyond the bounds of useful speculation. They make too many steps of inference without experimental check" (1949, 79). Of course, that was more than one-half century ago. Since Hebb's speculations, LTP has emerged as a promising candidate for experience-driven synaptic plasticity. LTP's story begins explicitly in Norway, 1973, although activity-driven synaptic efficacy had been reported anecdotally for at least one prior decade in the experimental electrophysiological literature. Three researchers in Per Andersen's laboratory—Timothy Bliss, Terje Lømo, and Anthony Gardner-Medwin—artificially stimulated fibers in the perforant path of both anesthetized and unanesthetized rabbits (reported in Bliss and Lømo 1973 and Bliss and Gardner-Medwin 1973, respectively). This fiber bundle is a collection of axons that synapse on granule cells in the dentate region of the hippocampus. Available electrophysiological techniques at that time enabled them to measure

- the amplitude of the population excitatory postsynaptic potential (EPSP) in granule cell dendrites that synapse with perforant path axons, indicating depolarization in granule cells generated by electrical stimulation of the perforant path;
- the amplitude of the population spike in granule cells, measured in the cell body layer of the dentate region, indicating discharge induced in granule cells by perforant stimulation;
- the latency of the population spike measured as time from perforant path stimulation until initial peak discharge.

These descriptions may be going beyond some readers' comprehension. And the "real neuroscience" example I'm describing in this chapter depends on the details of these and more recent experimental studies. So for the "neurophysiologically challenged," let's begin with some basic cellular neuroscience.

3.2 Some basic cellular neuroscience[10]

Neurons—nerve cells—constitute one of two tissues in nervous systems.[11] They possess a characteristic lipid bi-layer cell membrane that can propagate an electric charge down its length. This is due to two biophysical features. First, there is an unequal distribution of various ions inside and outside the cell membrane, most notably, positively charged sodium (Na^+), potassium (K^+), and calcium (Ca^{2+}), and negatively charged chloride (Cl^-) and various amino acids. Second, the activity of selectively permeable channel proteins permits some of these ions to flow across the cell membrane. Since the negatively-charged amino acids remain inside the cell (being too large to cross through open channels), and in the "resting" membrane state there are far more Na^+ ions in the extracellular space, a neuron's membrane has a negative resting charge or potential, typically around –70 millivolts (mV) (although this value has some variation across different neuron types). This potential is maintained by

1. the combined equilibrium potentials of the four key ions, and
2. active transport mechanisms.

The first is the membrane electric potential, for each type of ion, at which it has an equal probability of flowing across the membrane in either direction. The second are proteins that transport ions across the cell membrane *against* their diffusion gradients or electrostatic pressure. "Active" refers to the need for energy to drive this transport, derived from internal cellular sources. *Sodium-potassium pumps* are the most prominent of these in neurons. They actively transport Na^+ out of and K^+ into the neuron, despite greater concentration of Na^+ outside the cell (i.e., against its diffusion gradient), the negative resting potential inside the cell (i.e., against electrostatic pressure—"opposite charges attract, identical charges repel"), and the greater concentration of K^+ ions inside the cell (i.e., against its diffusion gradient).

Where we start to describe neuron physiology is somewhat arbitrary. We'll start at the *axon hillock*, with the generation of the *action potential* (or "spike") (Figure 2.2). In a typical neuron, this structure is located where the single axon emerges from the soma (cell body). The soma houses the cell nucleus, including the genes and gene expression machinery (more on this later) and the protein production machinery (outside the nucleus, in the cytoplasm). The axon is the neuron's "output line," through which it communicates with other neurons, muscle fibers, or other biological tissues. The axon hillock is dense with *voltage-gated ion channels*, especially ones selectively permeable to Na^+. Voltage-gated ion channels, proteins floating in the lipid bi-layer cell membrane, can change from closed to open with changes to the electric potential in nearby membrane patches. An ion channel is a configured protein. At resting membrane potential, its three-dimensional configuration prohibits the flow of ions across the cell membrane (although some leakage occurs regularly, in both directions: hence one need for Na^+-K^+ pumps, mentioned just above, to maintain the membrane's resting potential). But when the membrane potential in surrounding patches changes, an electric field is created that alters the configuration of these channels, opening them to ion influx or efflux. Channels selective for Na^+ are particularly important. When opened, Na^+ rushes into the neuron along its diffusion gradient and due to electrostatic pressure, since there is more Na^+ outside the cell and the inside is negatively charged relative to the outside. The net result is a sudden positive change in localized membrane potential.

If this potential exceeds the *threshold of excitation*, around –60mV in typical neurons, voltage-gated Na^+ channels open and the membrane potential in a local axon patch quickly "spikes" to around +50mV.[12] (See Figure 2.3.) About halfway through this electric "spike," voltage-gated K^+ channels (with a higher voltage threshold than Na^+ channels) begin opening. With the membrane potential now far above K^+'s equilibrium potential and a higher concentration by the forces of diffusion and electrostatic pressure. Within 1

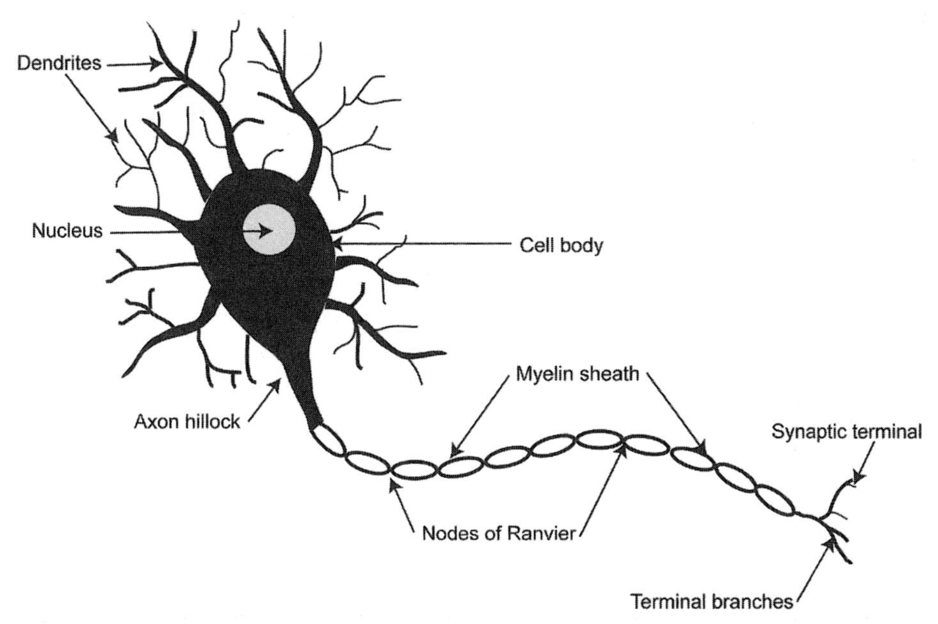

Figure 2.2. A schematic neuron, indicating various subcellular components. Figure created by Marica Bernstein.

millisecond (msec) (1/1000 of a second) after reaching their threshold of excitation, Na^+ channels begin to close. They undergo a brief refractory period (about 1 msec), during which no more Na^+ enters the cell. K^+ channels remain open, and K^+ efflux returns the local membrane potential back toward resting value (-70mV). By the time these channels close, the membrane potential has "overshot" its resting potential and become hyperpolarized due to the accumulation of K^+ outside of the membrane. Resting membrane potential is quickly re-established by the diffusion of K^+ throughout the extracellular fluid and the active transport of Na^+ ions out of and K^+ ions into the cell (the principal role of the Na^+-K^+ pumps). The entire process takes less than 3 msecs, after which that localized patch of axonal membrane is ready to "spike" again.

The special feature of axons is the *conductance* of action potentials down their entire length. The first basic "law" of axonal conductance is the *all-or-none law*. An action potential either occurs down an entire axon or it does not. Once generated in a localized patch, an action potential transmits down the axon to its end. An action potential remains the same size (same peak voltage value) during conduction. This is different from *decremental conduction* in passive electric *cables*, in which signal size decreases with distance from generating source owing to leakage and resistance. This

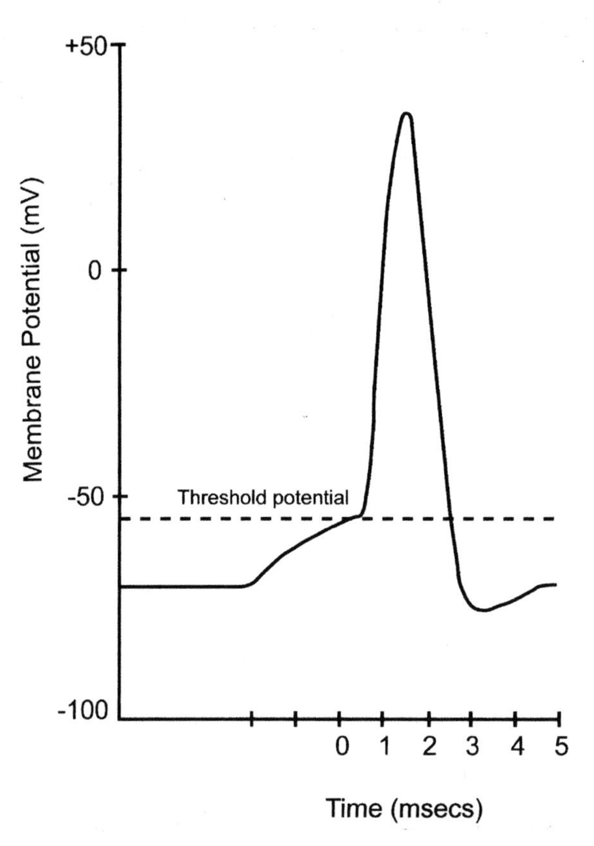

Figure 2.3. A schematic neuron action potential, graphed as a function of membrane potential to time. See text for a discussion of the action potential's biophysical mechanisms. Figure created by Marica Bernstein.

difference is due to the voltage-gated Na^+ channels that generate action potential depolarization. A localized depolarization creates an electric field that drives nearby axonal membrane patches over their thresholds of excitation, opening voltage-gated Na^+ channels and generating an action potential there. This occurs all the way down the length of the axon membrane.

Conductance of action potentials is facilitated by *myelination*, a feature of many (but not all) axons. Two types of glial cells—oligodendrocytes in the central nervous system and Schwann cells in the peripheral nervous system—wrap tightly around axons, leaving no extracellular fluid in between. Myelination leaves only small patches of axon membrane, at the *nodes of Ranvier*, in contact with extracellular fluid (see Figure 2.2 above). These are the only places on myelinated axons where Na^+ influxes through open voltage-gated Na^+ channels. Myelinated axons passively conduct electric current from node to node, exceeding the threshold of excitation to re-trigger an action potential at each. This so-called *salutatory conductance* has two advantages over continuous action potential conductance in unmyelinated neurons. First, it is less energy intensive. Active transport of Na^+ and K^+ to reestablish the action potential in each local patch requires a significant expenditure of a neuron's metabolic resources (measured up to 40% in some recent studies). Since Na^+ enters everywhere along an unmyelinated axon, these pumps must also be located all along the axon's length. In a myelinated axon, however, these pumps need only be located around the nodes of Ranvier. The second advantage is speed. Passive conductance depends on a cable's diameter. The thicker the cable, the faster is its passive conductance velocity. However, myelin serves as an insulator, which also increases conductance velocity. The fastest myelinated axon conducts action potentials at almost four times the velocity of the much larger squid giant axon, despite the latter's being 25 times the diameter of the former.

Action potentials in motor neurons that synapse directly on muscle fibers control the intensity of muscle contractions. Action potentials in sensory neurons reflect stimulus intensity. It is not a far stretch from these facts to the general idea, prominent in current neuroscience, that action potentials are the currency of neuronal information exchange. But if action potentials in individual neurons are all-or-none and constant in size, how can they represent variable parameters? The current idea is that variable information is represented by variable action potential (or "firing") *rates*. A higher firing rate—more action potentials per time unit—in a motor neuron produces a stronger muscle fiber contraction (by a mechanism we'll review momentarily). A stronger stimulus produces a higher action potential rate in selected sensory neurons. This has been dubbed the "rate law," which complements the all-or-none "law" (e.g., Carlson 1996, 40-41).

So far we've only sketched half of the story of neuron physiology, namely neural conductance. We turn now to the process of *neural transmission*. We left off with the all-or-nothing action potential reaching the end of the axon. What happens next? Axons typically branch into numerous *terminals* (see Figure 2.2 above). There the arriving action potential changes the configuration of voltage-dependent Ca^{2+} channels, opening them. Ca^{2+} influxes into the axon terminal by the forces of diffusion and electrostatic pressure. There is a far higher concentration of Ca^{2+} in the extracellular fluid, and despite the temporary depolarization of the membrane produced by the action potential, Ca^{2+}'s equilibrium potential is far higher than the membrane voltage at the peak of the action potential. Inside the terminal, Ca^{2+} binds with various proteins to bind the membranes of vesicles containing the neurotransmitter substance to the terminal membrane, producing a fusion pore through both that permits exocytosis of the transmitter substance. Released transmitter molecules diffuse passively into the *synaptic cleft*, the tiny space separating the *presynaptic* axon terminal from its *postsynaptic* target (see Figure 2.2 above). The postsynaptic cell, be it a neuron, a muscle fiber, or any other type of cell that synapses with neurons, contains *receptors*, proteins with a configuration that permits the binding of the transmitter molecule. Depending on the nature of the postsynaptic receptor, transmitter binding initiates a configurational change to the receptor protein that either initiates or inhibits electrochemical activity in the postsynaptic cell. If the postsynaptic target is a muscle fiber, transmitter binding either contracts the fiber or inhibits contraction by initiating well-understood biochemical activity. If the target is another neuron, transmitter binding either depolarizes or hyperpolarizes the membrane in the vicinity of the bound receptor.

Neurons contain two types of postsynaptic receptors. The simplest is the *ionotropic* or "fast" receptor. Transmitter binding alters the configuration of the receptor protein, opening a selective ion channel directly through it. Excitatory ionotropic receptors are selective for either Na^+ or Ca^{2+}. Ion influx (via forces of diffusion and electrostatic pressure) produces a small depolarization in the nearby membrane (far less than the threshold of excitation for generating an action potential). This occurrence is dubbed an *excitatory postsynaptic potential* (or *current*), or EPSP (EPSC). Inhibitory ionotropic receptors are selective for Cl^-. At resting membrane potential, open Cl^- channels have little effect, since the force of diffusion is offset somewhat by electrostatic pressure. There is more Cl^- outside the cell relative to inside, but resting membrane potential is very close to Cl^-'s equilibrium potential. When the membrane is depolarized, however, the forces of diffusion and electrostatic pressure drive Cl^- influx, hyperpolarizing the local membrane potential. This is an *inhibitory postsynaptic potential* (*current*), or IPSP (IPSC).

The other type of receptor, and by far the more prominent, is the *metabotropic* or "slow" receptor. Transmitter binding here does not directly open an ion channel. Instead, the reconfigured receptor protein activates intracellular G proteins. One subunit of an activated G protein (the α subunit) breaks away and binds to a distant ion channel, opening it to generate an EPSP or IPSP. Or, in an even more complicated scenario (which we'll see in detail later in this chapter), the α subunit binds to an enzyme protein, changing the latter's configuration to produce a *second messenger*: a protein product that translocates to other parts of the cell to initiate biochemical reactions there, including gene expression in the neuron's nucleus.

Following transmitter binding to postsynaptic receptor proteins, a number of events occur around and in the presynaptic terminal. Active transport mechanisms can *reuptake* transmitter molecules back into the terminal. This can happen either to the unaltered transmitter molecule or following its enzymatic degradation into inactive metabolites. Reuptake is a principal mechanism for terminating neurotransmitter actions, clearing the cleft for transmitter release driven by the next action potential. At high concentrations, transmitter molecules can diffuse through the extracellular fluid to bind to *autoreceptors*, protein receptors on the *presynaptic neuron* (not necessarily on the terminal). As a general rule, autoreceptors are part of negative feedback loops that slow or shut down transmitter synthesis or release by the presynaptic neuron. Glial cells also actively take up transmitter molecules from the cleft. Terminals from *modulatory neurons* can synapse onto primary presynaptic terminals (so-called *axo-axonal synapses*), releasing transmitters that bind to receptors on the presynaptic neuron to affect transmitter release (often through direct actions on action potential conductance or voltage-gated Ca^{2+} ions).[13]

On the postsynaptic side, EPSPs and IPSPs occur throughout the dendrites and soma. These local potentials interact along the membrane as they travel in both directions from their sites of origins. Interacting depolarizations and hyperpolarizations tend to cancel each other out, but similar charges summate both spatially and temporally. These are the processes of *neural integration*. When these interactions sum to the threshold of excitation at the axon hillock, an action potential is generated and propagated down the axon's length. Unlike the action potential, propagation down dendrites and soma is decremental. But owing to integration, the rate at which an axon fires—and hence represents information or controls muscle contractions—is determined by excitatory and inhibitory effects produced in its dendrites and soma.

Even the simplest spinal reflex involves thousands of interacting neurons—sensory, intraspinal, and motor. Thousands more are involved in even the simplest cortical inhibition of a spinal reflex. These numbers increase dramatically as we move to more complicated sensations, cognition, and behaviors. But every neuron involved is operating along the basic principles sketched in this subsection. These are, simply, the facts of the matter.

Let's pause here for a reductionist epiphany—in part to set the stage for what is coming, but also to remind readers why we are wading through neurobiological detail. If action potential rate is the currency of neural causation and information exchange, then *the only way an event can elicit neural change is by affecting the processes that underlie action potential generation in individual neurons*. That is where the rubber meets the road. Even those who champion "distributed" processing and "population coding patterns" must realize that neural populations are composed of individual neurons firing action potentials at variable rates. To affect, e.g., a population spiking frequency, an event must affect the individual neurons comprising the population. To do that, it must affect the processes governing action potential generation in individual neurons. In other words, it must affect the opening and closing of voltage-gated Na^+ and K^+ channels in the axonal membrane and the activity of Na^+-K^+ pumps that re-establish membrane potential, readying the axon for the next spike. Events that aren't "transducible" to that level of biochemistry and biophysics cannot affect neuronal activity. There are other "coding strategies" that are used in neuronal population studies besides frequency of action potentials.[14] But a population value of any sort is a function of that value in the individual neurons comprising it. And those neurons are nothing more than organized bags of molecular mechanisms for getting ions across selectively permeable membranes. As we will see as this chapter progresses, even pharmacological agents and the gene expression involved in synapse restructuring and formation, cell death, and so on, exert their effects on neurons at this level.

So you think, e.g., that poverty causes criminal behavior? Well, criminal behavior is (at the very least in part) a matter of orchestrated muscle fiber contractions. These contractions result from the release of specific molecules (acetylcholine, Ach) by motor neurons onto the endplates of individual skeletal muscle fibers comprising the affected muscle; and this release is controlled by differential spiking rates in individual motor neurons synapsing on specific fibers. A lot of cellular and molecular pathways affect motor neuron firing, many of which interact with other bodily systems. But all these features ultimately must affect the membrane proteins whose configurations at any given instant determine whether action potentials will be generated, and hence whether molecular transmitter substances will be released into neuromuscular junctions. I repeat the slogan I aired in the previous paragraph: for all their molecular biological and biochemical complexities, neurons are at bottom bags of processes that facilitate or hinder ions crossing selectively permeable membranes. So if, e.g., poverty causally influences behavior, it must be "transducible" down to this level of biochemical mechanism. This is causal-mechanistic "reductionism," minus philosophy of science's bells and whistles. Given what we know now about how neurons work (and my sketch so far has been at *the* most cursory level of cellular neuroscientific detail), if you deny this, you really are a causal dualist about behavior. "Not that there's anything wrong with that," to quote a repeated punch line from a famous *Seinfeld* episode, but theorists should own up to their commitments. If you reject a dualism of causal properties, for whatever reason, you cannot deny this implication.

3.3 Back to Norway, 1973[15]

Armed with this background (and its reductionistic implication), we return to Per Andersen's lab in Oslo, 1973. Bliss and Lømo (1973) inserted stimulating electrodes in the vicinity of axons forming the perforant path in anesthetized rabbits and recording electrodes into either the dendrite layer or the cell body layer of the hippocampus dentate gyrus. Perforant path axons synapse directly onto dendrites of dentate gyrus neurons. Conditioning trains delivered through the stimulating electrode consisted of either 10-20 electrical pulses/second for 10-15 seconds or 100 pulses/second delivered for 3-4 seconds. In 15 of their 18 rabbits, Bliss and Lømo (1973) found potentiated responses in dentate gyrus neurons to a single stimulating pulse delivered after a conditioning train, as compared to responses to the same pulse delivered prior to the conditioning train. This potentiation was measured as increased amplitudes in both the population EPSP measured in the dentate dendrite layer and the population spike measured in the dentate cell body layer, and decreased latency to population spike in the latter. These potentiations were measured from 30 minutes up to 10 hours after conditioning stimulus trains. All parameters were potentiated in nearly 30% of all experimental trials. Reduction in population spike latency was the most common result, potentiating in nearly 60% of all experimental trials. Amplitude of population EPSP increased in over 40% of all experimental trials, and amplitude of population spike potentiated in 40%.[16]

Bliss and Gardner-Medwin (1973) obtained similar results in a follow-up study using chronically implanted stimulating and recording electrodes in unanesthetized, alert, active rabbits. After single trains of conditioning stimulation at 15 pulses/second delivered via the stimulating electrode to a region of the perforant path, they measured long-lasting potentiation to single stimulating pulses on all three parameters on 41% of all experimental trials. Measurable potentiation lasted from 1 hour to 3 days in these preparations, and increasing the number of conditioning trains induced long-lasting facilitation for increasing durations. By demonstrating potentiation in unanesthetized animals, Bliss and Gardner-Medwin (1973) showed that the effect did not depend on neurons being in a depressed state.

Bliss and Lømo (1973) were quick to note potential implications for memory research. They remind us that "the perforant path is one of the main extrinsic inputs to the hippocampal formation, a region of the brain which has been much discussed in connection with learning and memory" (1973, 355). Clearly they were aware of lesion and neuropsychological research that had already linked the hippocampus to memory, as they cite two important review papers in their bibliography (Douglas 1967 and Olds 1972). In conclusion, they assert that "our experiments show that there exists at least one group of synapses in the hippocampus whose efficiency is influenced by activity which may have occurred several hours previously—a time scale long enough to be potentially useful for information storage" (1973, 355). Similar results by Bliss and Gardner-Medwin in awake, alert, mobile animals justified their conclusion that "it is at least possible that [the then-unknown mechanisms of this effect] could underlie some forms of plasticity under normal conditions in the hippocampus" (1973, 373). The cellular investigation of learning and memory now had a new focus.

Since this start, LTP has been the subject of much *in vitro* and *in vivo* research. Besides the memory-like characteristics that Bliss, Lømo, and Gardner-Medwin discovered, results have yielded quantifiable data about

- LTP's enhancement by repetition of conditioning trains (reminiscent of the consolidation effects of stimulus repetition and rehearsal).

Pharmacological and physiological manipulations have also revealed that LTP is both

- selectively blocked by treatments that inhibit certain types of long-term memory (measured behaviorally), and
- induced by physiological manipulations that augment types of long-term memory (measured behaviorally).

The experimental literature on LTP and its connection with behavioral work on memory, landmarked by excellent review papers, is a prime example of first-rate science.

4 MOLECULAR MECHANISMS OF LPT: ONE CURRENT MODEL

A recent account of LTP that links intracellular molecular mechanisms to measurable behavior is especially exciting. Over the past two decades, much cell-physiological research on LTP in mammals has shifted from the perforant pathway to the hippocampal Schaffer collateral pathway. The Schaffer collateral pathway is a bundle of axons from cells in the hippocampal CA3 region that project excitatory synapses to the hippocampal CA1 region. The connection between the hippocampus and long-term memory storage and access is now even more direct than it was in 1973. Bilateral hippocampal ablation in nonhuman primates produces little deficit in initial learning and short-term recall, but profound deficits on certain types of long-term recall tasks. It has been proposed as an animal model of human global amnesia, which also results from bilateral damage to hippocampus (and some surrounding tissue in the medial temporal lobe). Medial temporal lobe amnesics, like their experimental animal counterparts, have intact learning and short-term memory but profoundly deficient long-term memory for "declarative" items (Squire 1987).[17]

4.1 Early phase LTP

It is now common to distinguish distinct stages of LTP, based on their method of induction and temporal stability (Izquierdo and Medina 1997). *Early phase LTP* (E-LTP) begins immediately after a single high-frequency electric pulse train is delivered to Schaffer collateral fibers. Enhanced responses to subsequent pulses persist from one to three hours in CA1 neurons containing potentiated synapses. E-LTP induction and maintenance does not require gene expression or new protein synthesis. Nguyen *et al.* (1994) elegantly demonstrated this using hippocampal slices (400 microns thick) through the CA1 region from young (5-week old) rats. First they established baseline field EPSP response rates to Schaffer collateral stimulation prior to LTP induction. Then they induced LTP with three high frequency pulse trains through the stimulating electrode in both control and experimental slices. Immediately after these pulse trains ended, experimental slices were bathed with a solution containing a nonspecific gene transcription inhibitor, actinomycin D (ACT D). ACT D is a polypeptide antibiotic that infuses into DNA molecules, forming a stable drug-DNA complex. This inhibits DNA-mediated RNA polymerase activity, blocking transcription of messenger RNA, an early step in gene expression.[18] ACT D subfusion had no statistically significant effect on LTP for more than two hours nor on baseline field EPSPs for greater than three hours (Nguyen *et al.* 1994, Figure 1A, B). This result is especially interesting because it has been known for some time that short-term memory in behaving animals is also relatively impervious to gene transcription and protein synthesis inhibitors (Davis and Squire, 1984).

E-LTP induction involves two types of ionotropic ("fast") postsynaptic receptors for glutamate, the principal excitatory neurotransmitter in the mammalian central nervous system. Presynaptic terminals of Schaffer collateral fibers stimulated by the conditioning pulse train release increased amounts of glutamate due to the enhanced frequency of action potentials. Transmitter molecules bind to postsynaptic AMPA receptors (α-amino-3-hydroxy-5-methyl-4-isoxazole proprionic acid) (Figure 2.4A). Glutamate binding changes the configuration of the AMPA receptor protein, opening a channel through it selective for Na^+. Na^+ rushes into the cell through this opened gate via its diffusion gradient and electrostatic pressure, producing enhanced depolarization (positive current) of membrane potential in the vicinities of bound receptors/open Na^+ channels (all by the biophysical properties outlined in the previous section).

Enter next a second type of postsynaptic glutamate receptor, the NMDA (N-methyl-D-aspartate) receptor (see Figure 2.4A). At resting and weakly depolarized membrane potentials, NMDA receptors are blocked to glutamate's influence by voltage gated magnesium ions (Mg^{2+}) embedded within the protein's configuration. Under conditions of sufficient membrane depolarization, via activated AMPA receptors in the near vicinity, the Mg^{2+} block pops out, permitting glutamate binding. Bound glutamate changes the protein's configuration to open a direct channel selective both for Na^+ and Ca^{2+}. Ca^{2+} rushes into the cell via its diffusion gradient and electrostatic pressure. (Even though the membrane potential is positive due to nearby AMPA receptor activity, Ca^{2+}'s equilibrium potential exceeds even highly depolarized membrane potential.) At the same time, continued Na^+ influx through open AMPA receptors activates a family of intracellular enzymes, the tyrosine kinases, inside the postsynaptic terminal. These enzymes phosphorylate a subunit of the NMDA receptor protein, further enhancing the channel's Ca^{2+} conductance (see Figure 2.4B). Phosphorylation is a process by which a phosphate group (PO_4) gets attached to a protein. This changes the protein's three-dimensional configuration, leading to a variety of changes in the protein's interactions within the cell. Phosphorylation (and de-phosphorylation) is a crucial step in many cell-biological processes; it will come up often throughout the rest of this and the next chapter. In turn the increased Ca^{2+} levels in the postsynaptic cell set in motion a biochemical cascade involving numerous intracellular enzymes and interactions (Soderling and Derkach 2000). Ca^{2+} binds with calmodulin (CaM), increasing the intracellular level of the Ca^{2+}-CaM complex. This increase has two crucial effects. First, it stimulates adenylyl cyclase molecules that convert adenosine triphospate (ATP) into cyclic adenosine monophosphate (cAMP). Conversion of ATP into cAMP is a principal source of energy that drives cellular metabolic processes. But cells have also come to use the by-product of this process, cAMP, as a "second messenger" intracellular signal. (More on this below.) Second, increased levels of Ca^{2+}-CaM complex stimulate the autophosphorylation of Ca^{2+}/calmodulin-dependent protein kinase II (CaMKII) into its active form. Phosphorylated CaMKII (P-CaMKII) in turn interacts with the phosphorylated NMDA receptor, maintaining the open receptor's affinity for Ca^{2+} influx (see Figure 2.4B).

From here, two separate chains of molecular events occur within the postsynaptic terminal to maintain increased Na^+ conductivity through bound AMPA receptors. First, the increased levels of cAMP bind to regulatory

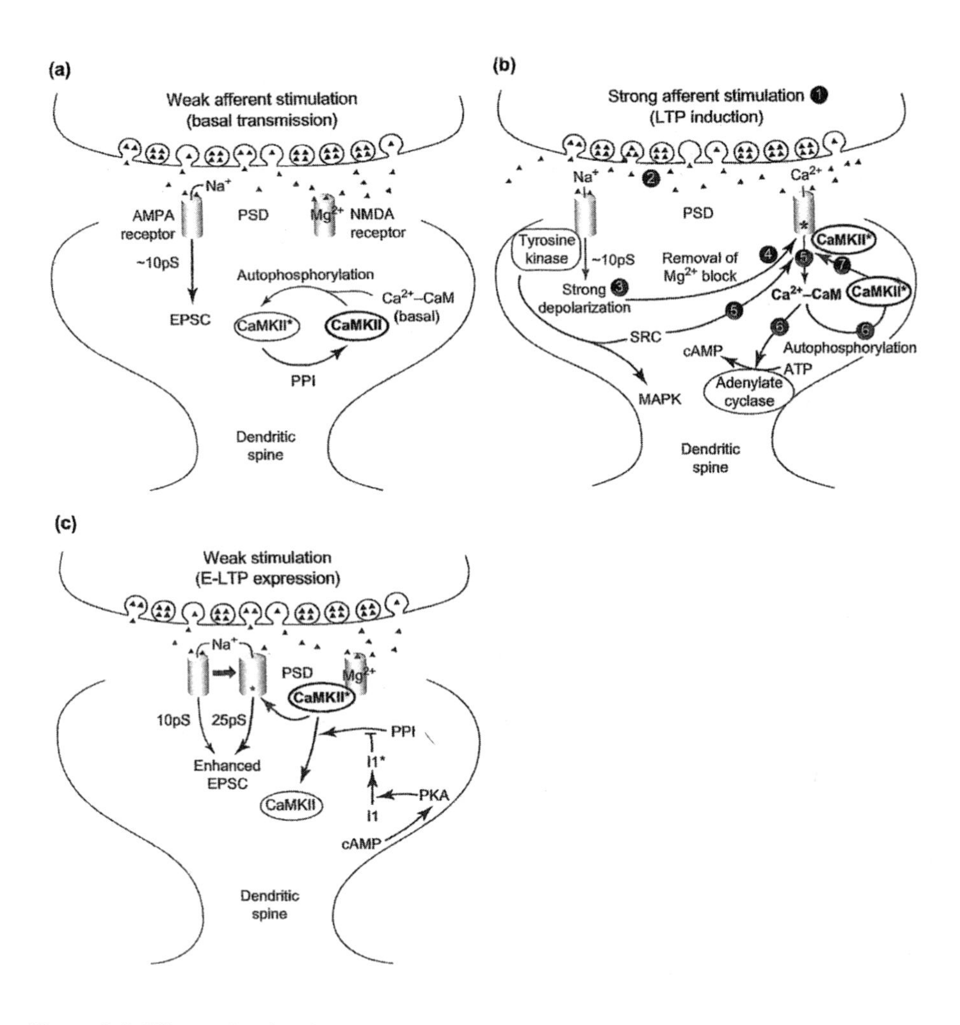

Figure 2.4. Effects of molecular mechanisms of E-LTP. A. Postsynaptic conductance capacity through bound AMPA receptors under conditions of a weak stimulation to presynaptic afferent and no potentiation. NMDA receptors remain under Mg^{2+} blockade. B. Strong afferent stimulation yields increased AMPA receptor binding, strong membrane depolarization, release of NMDA receptor blockade, glutamate binding to NMDA receptor, Ca^{2+} influx through open NMDA receptors, and an intracellular biochemical cascade that increases conversion of ATP into cAMP. C. Biochemical cascade yields AMPA receptors with a nearly three-fold increase in conductance capacity for Na^{+}. The weak presynaptic stimulation now yields an enhanced postsynaptic response. The synapse remains potentiated due to these molecular mechanisms for roughly one to three hours. See text for full explanation. Reprinted from *Trends in Neurosciences,* **23,** T. Soderling and V. Derkach, "Postsynaptic protein phosphorylation and LTP," Pages 75-80, Copyright (2000) with permission from Elsevier Science.

subunits of protein kinase A (PKA) molecules (see Figure 2.4B). PKA consists of two regulatory protein subunits and two catalytic subunits. When bound to the regulatory subunits, the catalytic subunits remain relatively inactive. But cAMP binding releases the PKA catalytic subunits, making them active in cellular interactions. First they phosphorylate inhibitor 1 (I1), which in turn inhibits the activity of protein phosphotase 1 (PP1). PP1 dephosphorylates both CaMKII and the subunit of the NMDA receptor, returning them to their inactive states. With this inhibitory enzyme inhibited, both CaMKII and the NMDA receptor subunit remain in their active (phosphorylated) state. Second, P-CaMKII in turn phosphorylates a subunit on the AMPA receptor, nearly tripling its conductance capacity for Na^+ (see Figure 2.4C above). The result is a temporary increase in the conductance capacity of AMPA receptors. While in this potentiated state, a weak stimulation to the presynaptic fiber—which in the basal state would produce only a weak depolarizing current in the postsynaptic membrane—now produces by itself an enhanced excitatory postsynaptic potential (EPSP) due to the amount of Na^+ that influxes through the potentiated AMPA receptors. This increased EPSP in turn raises the probability of spatial and temporal current summation at the axon hillock of the postsynaptic cell, and hence the probability that the postsynaptic cell will fire an action potential in distal response to presynaptic neurotransmitter release. This in turn raises the probability that its activity will influence action potential rates in the neurons it synapses on ... all the way to motor effectors that drive muscle contractions. (Bear in mind that LTP is occurring in selected synapses all the way down this cortical pathway.) Potentiated response in each affected synapse to weak presynaptic stimulation and glutamate release sets off the entire NMDA receptor-coupled intracellular biochemical cascade all over again, maintaining potentiation in each affected synapse in the circuit. These molecular mechanisms of E-LTP persist from one to three hours.

The molecular story of E-LTP grows even more interesting in light of recent discoveries about *retrograde neurotransmission*, from post- back to presynaptic terminals. The most prominent retrograde neurotransmitter of late is nitric oxide (NO). The biochemical details of its activity are still being unraveled, but the basic process and its influence on E-LTP are now experimentally verified in cultured Schaffer collateral-CA1 slices (Arancio *et al.* 1996, Son *et al.* 1996) (Figure 2.5). Increased Ca^{2+} influx through opened postsynaptic NMDA receptors activates NO synthase, an enzyme that stimulates NO production from the amino acid l-arginine. Being a soluble gas, NO readily diffuses out of the postsynaptic cell and into the presynaptic cell, where it has effects only at active terminals. There it activates soluble guanylyl cyclase and cyclic guanine monophospate-dependent protein kinase. This presynaptic cascade is known to be involved in enhancing neurotransmitter (glutamate) release (Arancio *et al.* 2000). So the molecular mechanisms of LTP not only make glutamate-binding postsynaptic AMPA receptors better conductors of Na^+ currents; they also generate a retrograde signal that enhances presynaptic glutamate release. The combined result is an even sharper increase in EPSPs at E-LTP affected synapses, persistent for up to 3 hours. This temporal feature correlates nicely with experimental psychological work on short-term memory, the labile, disruptable, transient form of virtually every type of memory (Squire and Kandel 1999, chapter 6).

4.2 Late Phase LTP

One remarkable feature of LTP is its temporal duration. Affected synapses can remain potentiated for hours to days. What molecular mechanisms "consolidate" and maintain LTP beyond the one-to-three hour duration of E-LTP? Inducing late phase LTP (L-LTP) in the Schaffer collateral pathway requires a *series* of conditioning pulse trains to the presynaptic fibers. Notice right away that this is a physiological laboratory analog of stimulus repetition (or rehearsal), known from psychological studies dating back over one century to be necessary for "consolidating" most short-term memories into stable, durable, long-term memories. In affected synapses, this increased presynaptic stimulation increases the rate and amount of glutamate released by presynaptic terminals, and in turn the number of bound postsynaptic AMPA receptors. The resulting increase in Na^+ influx further depolarizes the postsynaptic membrane, opening more voltage-gated NMDA Ca^{2+} channels and even more quickly raising the level of intracellular Ca^{2+}-CaM complex. Up to this point, the molecular mechanisms of L-LTP are identical to those of E-LTP; but next a new player arrives.

Modulatory interneurons become active, releasing a monoamine as their neurotransmitter—dopamine (DA), in mammals—and synapsing directly on CA1 dendritic spines in close proximity to bound AMPA and NMDA receptors (Figure 2.6). Their postsynaptic DA receptors are metabotropic. As described in the previous section, these receptors do not open ion channels directly, but instead engage the postsynaptic cell's metabolic machinery through a *second messenger*. DA binding activates a G-protein, guanine triphosphate (GTP), which couples to adenylyl cyclase to convert ATP into cAMP (the latter is the "second messenger") (Bernabeu *et al.* 1997). As with the molecular mechanisms of E-LTP, cAMP binds to the regulatory subunits of PKA, rapidly increasing the level of free catalytic PKA subunits in the postsynaptic cell. But this increase initiated by the second messenger generates enough freed catalytic PKA subunits so quickly that the latter translocate to the cell's nucleus and the "consolidation" events begin that distinguish L-LTP from E-LTP.

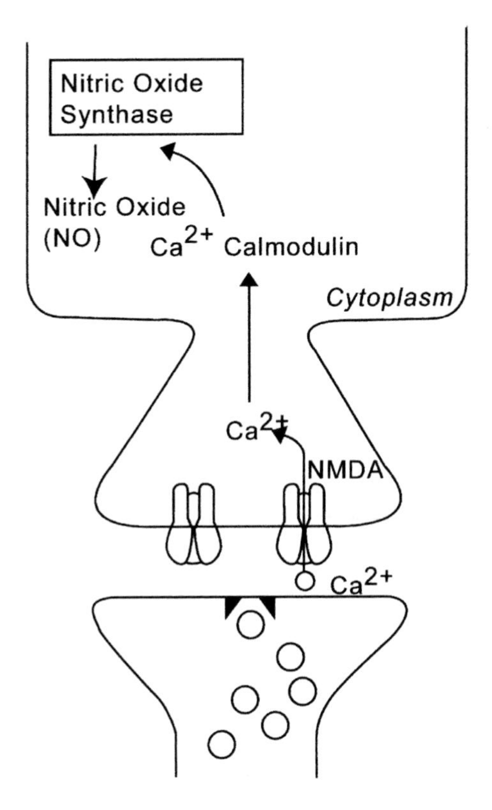

Figure 2.5. Mechanisms of postsynaptic nitric oxide (NO) production and retrograde transmission, driven by the molecular mechanisms of E-LTP induction. NO, a gas, diffuses across the cell membrane back into the presynaptic neuron, where it increases glutamate release rate through a G protein intracellular pathway. This increased glutamate release falls onto already-potentiated AMPA receptors. Reprinted from *Cell*, **88**, T. Abel *et al.*, "Genetic demonstration of a role for PKA in the late phase of LTP and in hippocampus-based long-term memory," pages 615-626, Copyright (1997), with permission from Elsevier Science.

As with the induction of long-term memory in behaving animals (Davis and Squire 1984), L-LTP induction requires new gene expression and protein synthesis. Consider again the Nguyen *et al.* (1994) study discussed in the previous subsection. Immediately after multiple pulse trains were deliv-ered through the stimulating electrode, hippocampal slices were dunked into either ACT D or a different transcriptional inhibitor, 5, 6-dichloro-1-β-D-ribofuranosyl benzimidazole (DRB) with a different mechanism of action on RNA polymerase activity. Activity following single pulses returned levels

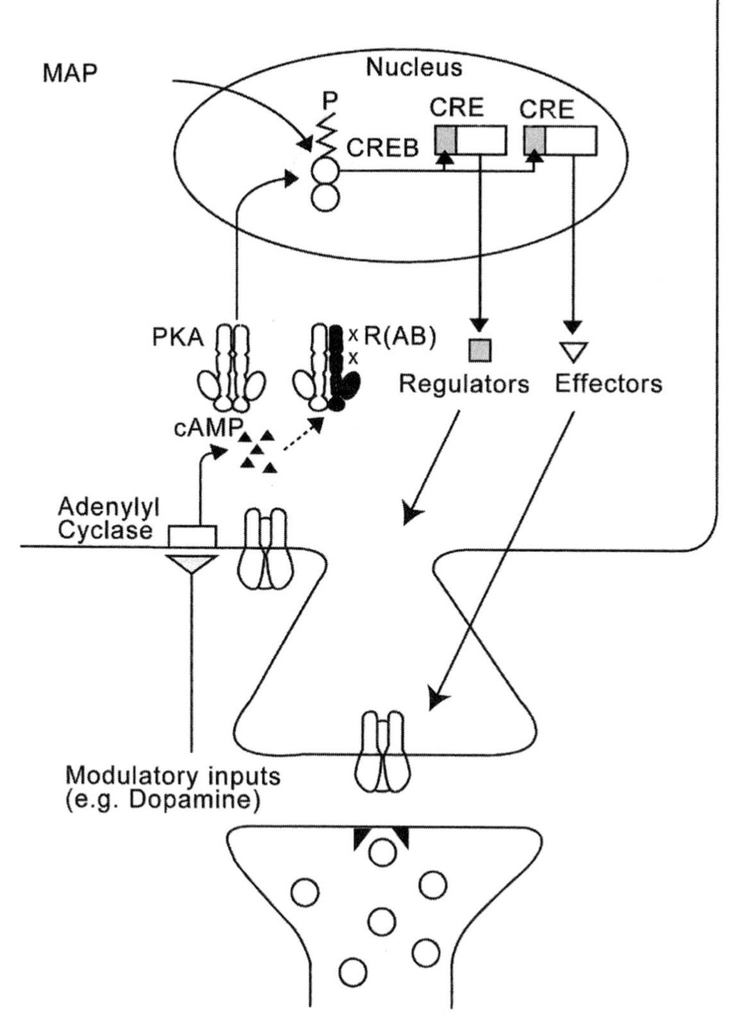

Figure 2.6. Early steps in the molecular mechanisms inducting L-LTP. Increased amount of cAMP molecules bind to the regulatory subunits of more PKA molecules, enabling freed catalytic PKA subunits to translocate to the neuron's nucleus. See text for detailed discussion. Reprinted from *Cell*, **88**, T. Abel *et al.*, "Genetic demonstration of a role for PKA in the late phase of LTP and in hippocampus-based long-term memory," pages 615-626, Copyright (1997), with permission from Elsevier Science.

within three hours. Control slices receiving the multiple pulse trains without either gene transcriptional inhibitor remained potentiated well beyond this time. Recall from the previous subsection that these nonspecific transcriptional inhibitors had no effect on E-LTP induction.

This experiment does not suggest which new genes and proteins are necessary for L-LTP induction; but one year earlier Frey *et al.* (1993) had published an experimental report that took its departure from previously published reports (mainly from Frey and his colleagues) that L-LTP was blocked by dopamine receptor D1 antagonists.[19] Since the bound D1 receptor was known to stimulate postsynaptic adenylyl cyclase, yielding an increase in cAMP and freed catalytic subunits of PKA, Frey *et al.* (1993) explored the hypothesis that L-LTP induction depended on PKA activity. They inhibited PKA activity in hippocampal Schaffer collateral-CA1 slices using Rp-cyclic adenosine 3', 5'-monophosphorothioate (Rp-cAMPS), which permeates the cell membrane and competes with cAMP for binding sites on PKA regulatory subunits. Once bound, however, Rp-cAMP does not release PKA catalytic subunits. Applying Rp-cAMPS fifteen minutes before inducing LTP had no significant effect on normal synaptic transmission for up to three hours and only a small negative (statistically significant) effect on E-LTP induction and maintenance 60 minutes after LTP-inducing stimulus trains. However, L-LTP in treated slices was completely blocked, measured as both field EPSP and population spike activities at three hours and longer after inducing stimuli (Frey *et al.* 1993, Fig. 1). The time course of L-LTP blockage by this selective PKA inhibitor closely matched those obtained using the nonselective protein synthesis inhibitors, suggesting that both results came from inhibiting the same mechanism.

They next applied a membrane-permeable cAMP analog, Sp-cyclic adenosine 3', 5'-monophosphorothioate (Sp-cAMPS), to hippocampal slices. This agent activates PKA. Even in the absence of conditioning pulse trains, this manipulation yielded an L-LTP time course similar to that induced in control slices in the usual fashion (via multiple pulse conditioning trains to Schaffer collateral fibers) (Frey *et* al.1993, Figure 3). A standard biochemical tetanization procedure further revealed that cAMP levels were elevated significantly one minute later in slices that received the three-pulse stimulation that induces L-LTP but not in slices that receive a single stimulating pulse, which only induces E-LTP. Elevated levels were not found in slices receiving the three-pulse stimulation along with SCH 23390, a D1 receptor antagonist, or with DL-2-amino-5-phosphonovaleric acid (DL-APV), an NMDA receptor antagonist. Ten minutes later, cAMP levels in slices receiving the three-pulse stimulating current and no inhibitory treatment had returned to normal.

Based on these results, Frey *et al.* (1993, 1663) conclude that L-LTP

- requires protein synthesis,
- seems to be maintained by the generation of cAMP and activation of PKA, and
- begins with PKA activity and new protein synthesis that starts within the first hour of LTP induction, while E-LTP is still in progress.

Viewed in retrospect, these studies were crucial steps toward one current model of the molecular mechanisms of L-LTP.

Infusing experimental techniques from molecular biology into mainstream neuroscience has since produced evidence of the specific gene expression and protein synthesis involved in L-LTP induction and maintenance. Catalytic PKA subunits, freed from their regulatory subunits by the rise in cAMP molecules, translocate to the postsynaptic cell nucleus. There they have two principal targets:

- they phosphorylate a class of *cAMP response element binding proteins* (the CREB-1 class), which in turn bind to *cAMP response elements* (CRE subregions) on regulatory regions of a variety of immediate early genes;
- they phosphorylate *mitogen-activated protein kinase* (MAP K), which in turn binds to a regulatory subunit on a second class of CREB proteins (CREB-2).

The overall effect is a changing balance in the molecular triggers that activate or inhibit new gene transcription and hence new protein synthesis. This molecular balance is sensitive to the intracellular biochemical cascades originating at synapses where LTP gets induced.

To understand these mechanisms we must make a brief foray (one paragraph and illustration!) into contemporary molecular genetics. Unfortunately, the rudiments of molecular genetics have been as ignored by philosophers of mind and many cognitive scientists as has been molecular neuroscience. (This dual neglect is not surprising since molecular neuroscience and molecular genetics fit together, hand in glove.) A gene can be divided into two functional regions: the coding region, whose base pair sequences code for DNA transcription into messenger RNAs (mRNA); and a control region, (typically) upstream to the coding region, whose base pair sequences in conjunction with regulatory ("response element binding") proteins control the initiation of the gene's transcription (Figure 2.7A). Transcribed mRNA molecules in turn translocate out of the cell nucleus and onto ribosomes in the cell cytoplasm for translation into new proteins. A gene's control region is further divided into *regulatory* and *promoter* regions (Figure 2.7A). An RNA polymerase enzyme binds to promoter proteins and the DNA in the gene's promoter region. Base pair sequences in the gene's regulatory region form response elements, to which regulatory protein molecules bind. Regulatory proteins come in two functional varieties: *response activators* and *repressors*. When response activators bind to response elements, this literally changes the shape of the DNA molecule, causing it to fold over and loop so that the regulatory proteins bound to it also contact and bind to the RNA polymerase (Figure 2.7B). This begins the process, or "turns on," transcription of mRNA. When either the appropriate activators are not bound to their response element or response repressors are bound there, the necessary alteration to DNA shape does not occur and gene transcription does not begin. Paired activators and repressors often compete for response element binding sites.[20]

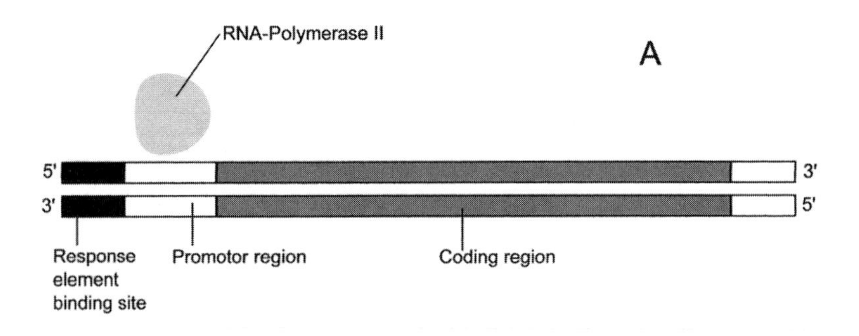

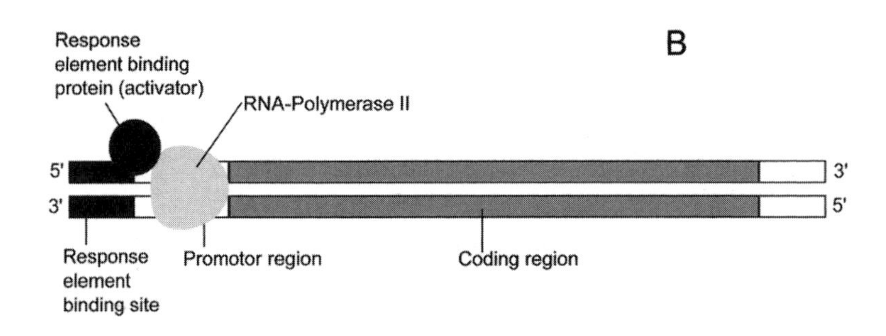

Figure 2.7. Basic constituents of a gene. A. No mRNA transcription initiated. B. mRNA transcription initiated by response element binding proteins bound to response element sites on the gene's regulatory region. See text for discussion. Figure created by Marica Bernstein.

We return to the molecular mechanisms of L-LTP. Phosphorylated cAMP response element binding protein (isoform) 1α (P-CREB-1α) is a response element binding *activator* for a variety of immediate early genes. When bound to a cAMP response element (CRE) site on these genes' control regions, it initiates gene transcription.[21] In L-LTP (in CA1 neurons), two types of *immediate early genes* are affected. Immediate early gene activation is a cell's earliest genetic response to a particular inducing stimulus. The protein products of immediate early genes typically act as enhancer binding proteins for gene expression further downstream, acting "immediately" to activate or repress transcription of other genes (in the same manner illustrated in Figure 2.7 above). In LTP induction, the inducing stimulus is the multiple-pulse conditioning train in afferent (input) fibers. Immediate early genes and their response profiles are common to cells of all tissue types, but are especially prominent (and well adapted) in neurons to support the latter's function as networked signaling units.[22]

One immediate early gene containing CRE sites to which P-CREB-1α binds transcribes mRNA to synthesize the protein, ubiquitin carboxyl-terminal hydrolase (Figure 2.8). This protein is part of an intracellular ubiquitin proteasome complex that destroys free regulatory subunits of PKA molecules. Without the activity of this complex, PKA regulatory subunits quickly bind back onto and re-inhibit the PKA catalytic subunits, halting the latter's synapse-potentiating effects (in both E- and L-LTP). But when these regulatory subunits are destroyed by the ubiquitin proteasome complex, intra-cellular PKA remains in a persistently active state, and so continues to translocate to the cell's nucleus and produce postsynaptic potentiation (Chain *et al.* 1999). Expression of the immediate early gene *uch* that transcribes ubiquitin hydrolase is thus driven by the very molecular process it in turn maintains.

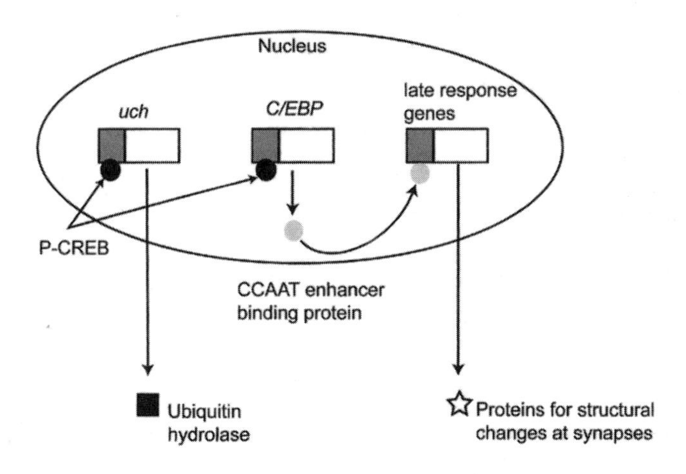

Figure 2.8. Specific gene targets of phosphorylated CREB. *uch* transcribes mRNA for ubiquitin hydrolase, a key constituent in a proteasome that destroys free PKA regulatory subunits, keeping the catalytic PKA subunits persistently active. *C/EBP* transcribes CCAAT enhancer binding protein, itself an activator for late-response genes transcribing mRNA for proteins that permanently potentiate postsynaptic structure and function. See text for details. Figure created by Marica Bernstein.

A second immediate early gene activated by P-CREB-1α codes for production of another transcription factor protein, CCAAT enhancer binding protein (C/EBP) (Figure 2.8). This protein is a response activator for a class of late response genes that code for proteins inducing growth of new postsynaptic dendritic spines, and hence new synapse formation (Taubenfield *et al.* 2001). The final result of these molecular mechanisms is a lasting increase in the number of synaptic sites between the presynaptic fiber and the postsynaptic neuron, a lasting structural change underwritten by new gene expression and protein synthesis. The increased number of synapses raises the probability of summated postsynaptic activity induced by presynaptic transmitter release. This heightened probability is, of course, L-LTP. Only now, the gene-driven structural changes persist for days or even longer, not just for a few hours.

These activational effects of P-CREB-1α are supplemented by phosphorylated mitogen activated protein kinase (MAP K), a second effect of translocated catalytic PKA subunits into the cell nucleus (Cammarota *et al.* 2000). A second type of CREB protein, CREB-2, is a repressor of the immediate early genes for which P-CREB-1α is an activator. The two molecules compete for CRE binding sites on the genes' control regions. CREB-2 lacks a binding site for catalytic PKA. But it has a binding site for P-MAP K, which in turn phosphorylates CREB-2 to inhibit its repressor activity. Shutting down CREB-2 further enhances the activation effects of P-CREB-1α. The interplay of this activator and repressor determines the ease or difficulty with which E-LTP "consolidates" into L-LTP.

I end this section with one final scientific question. How can this account of L-LTP's molecular mechanisms explain the *synapse specificity* of LTP? If L-LTP ultimately is governed by events occurring in the postsynaptic neuron's nucleus, how do the protein products "know" which synapses to translocate to and potentiate, and which to avoid? Given that potentiation is specific to active synapses, the idea of a stimulation-induced "mark" or "tag" seems theoretically promising. Such a tag could enable active synapses to capture and incorporate functionally the products of gene expression that have been transported nonspecifically throughout the cell. Recent empirical studies on invertebrate neurons have revealed components of gene translational machinery, including mRNA and ribosomes, in distal dendrites. The intriguing idea that active synapses synthesize "locally" their own protein "tags" that interact with and "capture" gene expression protein products being transported down from the soma has some recent experimental support (Martin *et al.* 1997), and is currently a topic of intense investigation.

The conditioning pulse trains that generate L-LTP eventuate in postsynaptic neurons whose gene expression, protein production, synaptic structure and biochemistry—and so their capacity to generate action potentials in response to afferent input—are *lastingly* altered. These same events are occurring at specific synapses throughout the neuronal circuitry leading from sensory receptors to motor effectors, namely, at the ones selectively activated by the afferent stimuli or endogenous neural activity. In the behaving animal, these molecular mechanisms eventuate in permanently changed motor responses (behavior) to specific sensory inputs or endogenous activity. Our next task is to see how recent neuroscience has built an explanation of behavioral data in memory consolidation tasks out of this developed account of the molecular mechanisms of E- and L-LTP.

5 BUT IS THIS REALLY MEMORY (CONSOLIDATION)?

I've imposed upon readers a fair amount of cellular and molecular detail and more is coming. So now is a good spot to remind you about where this discussion is headed and why these details are important. Given the induction similarities and temporal correlations between short-term memory, long-term memory, E-LTP, and L-LTP, a tempting hypothesis is that psychology's "consolidation switch" reduces to a three-part sequence described within contemporary cellular and molecular neuroscience. This sequence starts with activity-induced enhancement of freed catalytic PKA subunits at active synapses in specific neurons, leading to

1. increased binding of P-CREB-1α to CRE regulatory subregions on one immediate early gene that transcribes a protein for maintaining persistently active PKA, and another that transcribes a response element binding activator that turns on expression in late-response genes coding for proteins that generate the growth of new synaptic sites;
2. inhibition of CREB-2, a transcription repressor for the same immediate early genes; and
3. the translation of protein products transcribed by these immediate early and late-response genes.

What psychologists call "retrograde interference" turns out to be any process initiated after initial stimuli that disrupts any of these cellular/molecular steps. In Duncan's (1949) study (this chapter, section 2 above), electro-convulsive shocks delivered fifteen minutes or less after the unconditioned stimulus were very disruptive of these steps, while retrograde hind leg shocks were not.

Tempting as this hypothesis is, the key question remains outstanding: Can this potential reduction be verified experimentally? The most straightforward way to do this would be to manipulate these molecular mechanisms of L-LTP directly to produce observable effects that spare short-term memory but compromise long-term memory *behavior*. This result would be especially compelling if it involved a form of long-term memory specific to "higher" mammals, for which it is widely agreed that this form is "genuinely cognitive." This procedure has become central to the *explanatory* claims of these molecular mechanisms in recent mainstream neuroscience, and hence to the proposed *reduction* of mind to molecules. We now shift to the additional scientific background necessary to understand the experimental details that have recently yielded exactly these results.

5.1 Declarative memory

Declarative memory is just such a form. It admits of the short-term memory/long-term memory distinction and is specific to mammals. This term is best known from the work of neuropsychologist Larry Squire and his colleagues. Its origins lie in different patterns of learning and memory preserved versus lost in human *global amnesia*. The principal neuropsychological symptom of global amnesia is impairment to remembering new information, including acquisition, retention, and/or retrieval. Most human global amnesia occurs as part of Korskoff's syndrome, due to long-term alcohol abuse. But there are also selective cases in which the deficit results from bilateral damage to the medial temporal lobes, a part of the limbic system housing the hippocampus. Celebrated case H.M. is one of the latter. H.M. underwent surgical bilateral ablation of the medial temporal lobe in 1957 to treat otherwise intractable epilepsy. The surgery produced a profound memory impairment against a background of retained cognitive and intellectual capacities (Scoville and Milner 1957). H.M.'s anterograde amnesia for events that have happened since the surgery is virtually complete, and his retrograde amnesia stretches back for a few years prior to the surgery. Interestingly, his memory for events in his distant past, well before the surgery, remains intact, comparable to nonamnesics controlled for age. Up to the present day, he forgets events soon after they end. For example, he is unable to remember what he ate at his previous meal, he can't remember having met individuals before that he sees regularly (e.g., doctors, nurses), and he can't remember being told before that his parents have died. Nevertheless, soon after his surgery, neuropsychologist Brenda Milner showed that H.M. could acquire and perform a motor skill at a normal rate (compared with nonamnesics), despite his inability to recall prior episodes of training. The skill was pursuit-rotor, where subjects guide a mechanical device to keep it on target with a dot rotating on a turntable at variable speeds. Following this initial demonstration of retained skill acquisition capacity, neuropsychologists began to explore the extent of retained learning and memory capacities in global amnesics.[23]

Squire's distinction between declarative and nondeclarative memory systems emerged directly from this research. Figure 2.9 illustrates his current manner of carving up memory systems. The declarative system is compromised in global amnesics, particularly in medial temporal lobe amnesics. Squire and his collaborators initially chose the term, "declarative," to suggest the ability to declare one's knowledge verbally. This memory type or system has often been characterized in terms of conscious recognition and recall. In a fairly recent publication, for example, Squire introduces the term this way:

> One kind of memory provides the basis for conscious recollection of facts and events. This is the kind of memory that is usually meant when the terms 'memory' and 'remembering' are used in ordinary language. Fact-and-event memory refers to memory for words, scenes, faces, and stories, and is assessed by conventional tests of recall and recognition. This kind of memory was termed "declarative" to signify that it can be brought to mind and that its contents can be "declared." (1992, 232)

Clearly, this characterization picks out a type of memory that qualifies as "genuinely cognitive." The problem is that under this description it is difficult to apply to memory research using experimental animals. Connecting up the human neuropsychological literature with experimental animal work requires a broader characterization of declarative memory.

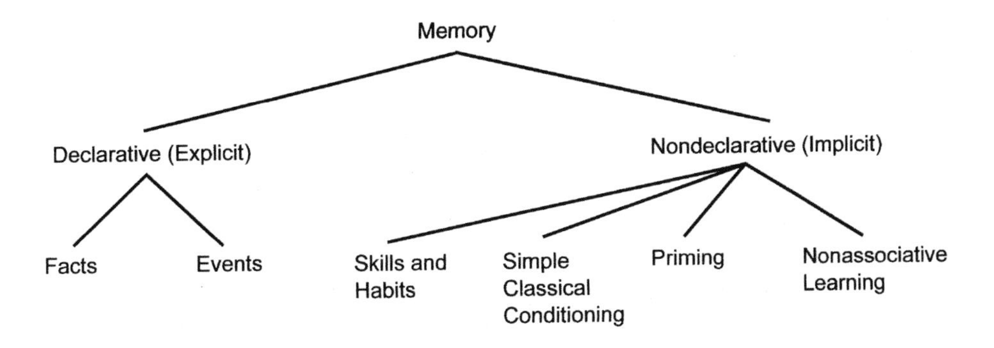

Figure 2.9. Squire and colleagues' current division of memory systems and subsystems. (See Squire 1992, 233.)

The key to achieving this, according to Squire, is to adopt a "brain systems framework" (1992, 240). No matter which of numerous popular terms one favors to denote the memory type or system compromised in human global amnesia—'declarative,' 'explicit,' 'relational,' 'configural'—they all converge on tasks that require an intact hippocampus and surrounding tissue in the medial temporal lobes. Squire writes:

> If one considers the various biological and purely psychological concepts that have been used, it is striking that they sort themselves out in terms of ideas about what the hippocampus does and does not do in the service of memory. ... The important point is that the terms *explicit* memory and *declarative* memory, when one considers the properties that have been associated with each, describe a biologically real component of memory that depends on particular structures and connections in the brain. (1992b, 205)

This approach forges a connection between human neuropsychological data and experimental mammalian research. The "particular structures and connections," namely the hippocampus proper, entorhinal cortex, perirhinal cortex, and perihippocampal gyrus, have homologs across the mammalian class. Since declarative (or explicit) memory is coextensive with hippocampal-requiring memory, the term is applicable to memory research on humans, other primates, and rodents (Squire 1992b).

Another result that holds across the mammalian class is that a second prominent structure in the medial temporal lobe, the amygdala, is *not* associated with declarative memory. Although H.M.'s surgical ablation included his amygdala bilaterally, humans with more circumscribed lesions that do not include it still display H.M.'s pattern of preserved and impaired memory capacities. Controlled studies with surgically lesioned monkeys indicate that hippocampal lesions (and surrounding tissue) produce significant deficits in delayed nonmatching to sample experiments, while amygdala lesions alone do not; and amygdala lesions conjoined with hippocampal lesions do not exacerbate this behaviorally measurable deficit (Squire and Zola Morgan 1991). In delayed nonmatching to sample, subjects are presented with an object and then, following a delay period, are presented with the original object and a novel object. They must choose the novel object (the "nonmatch to sample") to receive the task reward. Human global amnesics are impaired on this task. Results using it with surgically lesioned monkeys were crucial in exploring the extent of medial temporal lesions that interrupt declarative memory and establishing the currently accepted primate model of human global amnesia (Squire and Zola Morgan 1991).

Two procedures have been prominent in the rodent experimental literature on declarative/relational/hippocampal-requiring memory. One is the multiple-armed radial maze. This consists of a raised platform with a compartment in the center and hollow arms radiating out (like spokes of a wheel). Arms are baited with a small food reward at their ends away from the center compartment. Rats can be trained to search efficiently for food rewards; after around twenty trials, they never return during an individual trial to a compartment they visited earlier in that trial. They can even do so when prevented from following a fixed sequence of arm visits (e.g., when they are blocked from visiting next the arm immediately to the left). Rats with lesions to hippocampus, fornix, or entorhinal cortex, however, cannot learn to visit the arms efficiently. Whether or not they re-enter an arm later in a trial that they visited earlier in that trial falls to chance. On each trial they eventually find all the food, but only after entering many of the arms multiple times (Olton 1983). This is not due to an inability of the lesioned rats to distinguish among the arms. They can learn whether a given arm contains food across trials or never does. Olton and Papas (1979) demonstrated this result elegantly using a seventeen-arm radial maze. Before each trial, eight arms were baited; the other nine arms were never baited on any trial. Trained rats with fornix lesions visited the baited arms randomly and inefficiently, re-entering arms they had previously entered earlier in the trail. But they also learn at a rate comparable to controls to avoid the arms that are never baited. Apparently they cannot remember where they have visited recently, but they can remember which locations regularly contain food and which don't.

A second common experimental procedure is the Morris water maze. Rats and mice are hydrophobic; they don't like being in water. The Morris water maze takes advantage of this motivational fact. It is a pool filled with opaque liquid (originally water mixed with powdered milk) with a single submerged platform high enough for the rat or mouse to be perched mostly out of the water. Visual stimuli are painted on the walls of the maze (or the room containing it) so that the rodent can navigate a path to the platform. Rats are put in the pool initially at a random location and swim until they encounter the platform. They are then put back into the maze at a different location. Rats quickly learn to navigate to the submerged platform and after a few trials will swim directly to it, no matter where they are placed into the pool. In his original study, Morris (1982) found that rats with neocortical lesions performed similarly to normal controls, but rats with hippocampal lesions swim randomly on each trial until by chance they come upon the platform. This deficiency in memory for spatial locations is consonant with a similar deficiency in human amnesics. It falls clearly within the declarative/ relational/hippocampus-requiring category. However, hippocampal-lesioned rats and mice can learn and remember other tasks in the water maze—

nonrelational, stimulus-response tasks. For example, if the platform is elevated to just above water level, making it visually accessible, hippocampal-lesioned rats quickly learn to swim directly to it. Even with the submerged platform, if hippocampal-lesioned rats are always released at the same place in the pool, they learn to swim in the same direction every time.

However, the popularity of radial arm and water maze tasks sidetracked animal researchers into thinking that the rodent hippocampus was involved primarily in spatial memory. Recently, Howard Eichenbaum and his colleagues have showed that hippocampal-lesioned rats also fail on a nonspatial relational memory task. Bunsey and Eichenbaum (1996) trained rats to dig for food rewards buried in jars under scented sand. The odors were originally presented as paired associates. For example, on Training Set 1, the rats learn that if odor A is presented as the sample, they are to dig in the jar under the sand scented with odor B rather than odor Y. If odor X is the sample, then they are to dig in the jar under the sand scented with odor Y rather than odor B. After mastering that training set, the rats are presented with Training Set 2. Here they learn to pair sample odor B (from Set 1) with the choice of odor C instead of odor Z, and to pair odor Y (also from Set 1) with the choice of odor Z instead of odor C. (See Figure 2.10) Hippocampal lesioned rats acquire all paired associate combinations at a rate statistically similar to control sham rats (who undergo a similar surgical procedure leading up to the hippocampal lesions, but don't receive the lesions). Next Bunsey and Eichenbaum (1996) subjected the trained rats (hippocampal lesioned and sham controls) to a Test for Transitivity. Either odor A or odor X (both from Set 1) is presented as sample, followed by a choice between odors C and Z (both from Set 2). If A is the sample odor, the food reward is under C; if X is the sample, the reward is under Z (Figure 2.10). Sham rats performed far above chance in this test requiring a transitive operation over the paired associate odor memories; hippocampal lesioned rats, who learned the paired associates with equal ease, performed slightly below chance. Bunsey and Eichenbaum (1996) also tested the paired associate-trained rats on a Test for Symmetry. Odors are the same as in Training Set 2 of the transitivity study, but now samples are either odor C or Z and the choices are odors B and Y. If C is the sample, the food reward is buried under odor B; If Z is the sample, it is buried under Y (Figure 2.10). Here again, sham rats performed far above chance, whereas hippocampal lesioned rats were virtually at chance. These results are also consistent with reports of stimulus-stimulus association learning deficits in hippocampal lesioned monkeys (and surrounding tissue) and amnesic humans.

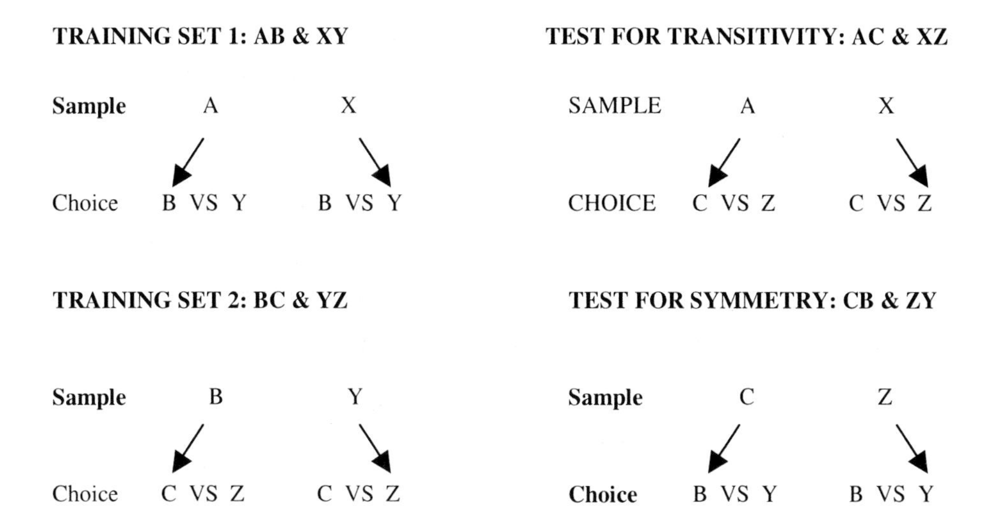

Figure 2.10. Odor paired associates in Bunsey and Eichenbaum's transitivity and symmetry memory tasks. (See Bunsey and Eichenbaum 1996, 256.)

Clearly, the hippocampus mediates more than just spatial learning even in rodents. Bunsey and Eichenbaum (1996) refer to this declarative/explicit/relational/hippocampal-requiring memory feature as *flexibility*, referring to the animal's ability to use such memories to solve novel problems other than the ones on which they were acquired. This is in contrast to the inflexibility of nondeclarative memories, which are tied directly to the stimuli or training regimen generating them. Cohen and Eichenbaum's (1993) picturesque phrase, the "promiscuity" of declarative memory, is also intended to display its accessibility by numerous and novel expressive routes. This feature locates declarative (/explicit/...) memory squarely within the realm of the "genuinely cognitive." And once again, in keeping with Squire's "brain systems framework" for memory, the common link underlying flexible, promiscuous memories across mammalian species is their dependence on the intact hippocampus (and surrounding tissue).

5.2 Biotechnology solves a long-standing methodological problem in LTP-memory research[24]

Now we can make more specific the question raised at the beginning of this section. Can a direct experimental link be established between the molecular mechanisms of E-LTP and L-LTP and the consolidation of *declarative* short-term memories into long-term memories? The search for experimental links between LTP and memory has been both exciting and frustrating—and continually subject to methodological criticisms. Many critics have focused on the pharmacological agents and procedures that up to a few years ago were the best available laboratory techniques (see, e.g., McEachern and Shaw 1996). Two limitations on pharmacological manipulations are obvious. First, pharmacological agents are rarely completely specific for a single type of protein target (receptor or enzyme). So even though elaborate delivery techniques can restrict application of an agent to an extremely local area, it can still be difficult to control for effects on molecules and proteins that the experimenter is not targeting. Second, and specific to the emerging story of the molecular mechanisms of LTP, as of the mid-1990s pharmacological manipulations had not yet revealed the relation between L-LTP and long-term memory in behaving animals. It was then that some groups turned to *targeted gene manipulations* that, owing to advances in biotechnology, can deliver maximally specific enhancement or disruption of particular molecules in intracellular pathways. A particular gene directs the production of a specific protein that helps to determine the structure or function of each cell in which it is expressed. By the early 1990s, techniques existed to manipulate any gene in the mouse genome.

Two of these techniques have dominated recent neuroscientific research. One generates "knock-out" organisms; the other generates "transgenics." Knock-outs show "what happens when we create deliberate mutations in a particular gene in a living organism," while "transgenics illuminate the role of a protein product's function in intracellular processes" (Weaver 1999, 127). Lodish *et al.* (2000) make an even grander claim about the importance of these techniques, opening their chapter describing them by asserting: "The goal of modern molecular cell biology is nothing short of understanding the biochemical, cellular and organismal functions of all the proteins encoded in the genome" (281). Interestingly, they end that discussion by describing "one important example ... [that] comes from studies on learning and memory" (2000, 286).

Ultimately, both knock-out and transgene techniques depend upon inserting foreign DNA into an organism's genome. Both techniques start with the same procedures. Since these technologies have been applied to many different research questions, molecular cell biologists have developed many technical variations. In this subsection we will describe the processes of making knock-out and transgenic mice in only general terms. But we will include enough detail to foster an appreciation for just how laborious (and expensive!) the technologies are, and for how rigorous the bench science must be.

If one intends to understand a gene's or protein product's biological function, then not just any nucleotide sequence can be inserted into the genome. The gene's specific nucleotide sequence must first be known, the sequence of adenine, thymine, cytosine, and guanine (the As, Ts, Cs, and Gs) that carry the DNA's information. Thus the first step in both knock-out and transgenic technologies is to identify and sequence the gene of interest (or its protein product). There are now several technologies that facilitate this, especially the creation of searchable "libraries" of organisms' genomes. Theoretically, this process is straightforward. One suspects that an expressed protein is important in a particular biochemical pathway or process occurring in a particular tissue at a particular time. One isolates the immediate product of DNA transcription, messenger RNA (mRNA), from those tissues.[25] The initial product will be mRNA molecules corresponding to all the genes being actively transcribed in those cells at that time. In order to sequence the gene of interest, its mRNA must be separated from the rest. Relatively large quantities of the initial product are needed for this separation (and for subsequent steps). Additionally it is DNA, not RNA, that will be inserted back into the genome, so the gene-specific mRNA sequence must be re-converted into a DNA sequence. Increasing the number of molecules (amplification) and converting mRNA into DNA is accomplished using the workhorse of molecular cell biology, the polymerase chain reaction (PCR). PCR is an *in vitro* biochemical reaction that exponentially increases the number of RNA or DNA molecules. Because the enzymes that make RNA copies of DNA in cells are incredibly faithful, they can be used in a special kind of PCR, reverse transcriptase-PCR (RT-PCR). This process both "back-tracks" mRNA into DNA and generates thousands of copies of cDNA ("complementary DNA," a DNA copy of an RNA molecule). From those cDNAs, the gene of interest is isolated and sequenced using any of a number of standard biochemical techniques. Most techniques filter the cDNA through an agarous gel that separates molecules according to their molecular weights. Individual "bands" on the gel, formed by deposits of many cDNA fragments of a given molecular weight, can be physically cut out of the gel and re-amplified. But appreciate that in order to isolate the right band, some features of the gene must already be known.

Note that the preceding chain of events need not occur in precisely this order, and that details depend on the specific objectives of the research. For our purposes, the important point is that the product will now be purified nucleotide sequences corresponding to the gene of interest. This cDNA can now be altered—mutated—in a variety of ways that render the cDNA or its protein product non-functional when inserted into a cell's genome. This is the essence of a "knock-out" gene. An endogenous functional gene is replaced by another that has been mutated in a specific way. However, as we saw in the previous section, not all regions of a gene wind up translated into a protein product. "Control" regions regulate transcription timing and rate (see Figure 2.7 above). If those regions are selectively mutated, then when the "transgene" is inserted into the genome, the resulting protein is functional but expressed in abnormal quantities.

Now the fun begins! The ultimate goal is to insert a DNA sequence into double stranded DNA in the nucleus of a mammalian cell. Fortunately for molecular biologists, nature has been inserting foreign DNA into native DNA for millennia, outside the laboratory. Viruses, phages, plasmids and even modified yeast chromosomes all have the capacity, when introduced into a host cell, to insert their own small genomes (or portions of them) into those of their hosts. Molecular biologists make use of plasmids as "vectors" to transfer and insert foreign DNA into mouse genomes. Plasmids are small circular pieces of bacterial DNA that replicate independently of the bacterial cell's chromosome. They also contain "restriction enzyme recognition sites" within their own nucleotide sequences. Restriction enzymes recognize relatively short but very specific DNA nucleotide sequences and break the phosphodiester backbone of DNA between them. They literally cut up DNA. There are many restriction enzymes found naturally in bacterial cells. There they function to remove foreign, potentially damaging DNA from bacterial genomes. By selecting the appropriate plasmid and restriction enzymes, molecular biologists cut the plasmid DNA and insert the mutated gene of interest directly into it. Other enzymes seal the break and we now have a circular piece of DNA, one portion containing the altered gene of interest. By asexual reproduction, these plasmids can replicate virtually indefinitely, generating many "clones" of the original gene.

Two additional details require mention. First, some biochemical tinkering with the gene is necessary so that the ends of its sequence are complimentary to the ends of the broken plasmid DNA. At this stage, the gene of interest is a single stranded DNA molecule. Single strands form double strands in virtue of complimentary base pairing: A-T, C-G. The sequence of one end of the gene must be engineered such that it is complimentary to one of the plasmid ends that results when the circular plasmid is cut by the restriction enzyme. In principle this is not difficult, since we know the restriction enzyme recognition site. Second, often the unaltered gene is inserted and the mutation is induced after the gene is in the plasmid. This method is useful when the mutation involves removing longer nucleotide sequences that might correspond to functional domains of the protein.

Amplifying, identifying, sequencing, and cloning are virtually identical for both knock-out and transgenic techniques, although subtle differences will depend on specific research objectives. At this point, two questions become critical:

1. Did the mutated gene of interest insert into the mouse's genome?
2. If so, where in the genome is the insertion located?

Knock-out and transgenic techniques differ in their answers to question (2). When creating the plasmid vector carrying the altered gene of interest, steps are taken to answer these questions. We will first focus on knock-out mice and return later to the construction of transgenics.

Remember that in a "knock-out" the functional wild-type gene is replaced by a deliberate mutation. Cultured mammalian cells are naturally permeable to DNA and there are several ways to increase this permeability. Thus the plasmid vector can be inserted into embryonic stem (ES) cells from brown mice. (As we will see below, coat color is useful to determine whether the mutated gene has been inserted properly.) Under the right conditions, some of those plasmids will translocate into the ES cell nucleus. Because ES cells are actively dividing, the enzyme machinery that promotes incorporation of the gene of interest into the genome is present and active. Gene insertion takes place via *recombination*, the physical exchange of DNA between two chromosomes, which is occurring naturally in reproducing cells. "Homologous recombination" is the actual "crossing over" of the two sister chromosomes during meiosis I, the stage of the cell cycle during which homologous chromosomes are closely aligned. During crossing over, a gene from the maternal chromosome is exchanged for the same gene on the paternal chromosome. If that recombination occurs in germ line cells, the resulting gametes will have genomes fundamentally different from both the parental genomes. That recombination occurs in sexually reproducing organisms explains in large part the huge amount of variation in populations; glibly, it explains why siblings can have such different combinations of their parents' features. The event that leads to insertion of foreign DNA is a recombination event.

Three things can happen to the foreign DNA in plasmid vectors entering the ES cell nucleus:

1. Nothing. No recombination takes place and no DNA is inserted.
2. Random recombination takes place in all or many chromosomes. The gene of interest is inserted, but only along with portions of the plasmid sequence and not at its proper location.

Or

3. homologous recombination occurs. The mutated gene of interest crosses over with a chromosome where the wild type gene resides. The wild-type is replaced by the gene of interest.

But how can we tell which of these three possibilities occurred?

We said earlier that steps are taken when constructing the plasmid vector to answer this question. Typically, two other genes are inserted into the vector along with the gene of interest. The first gene permits ES cells containing no insertions to be separated from those containing some insertions. The classic technique is to insert a gene that confers resistance to an antibiotic, usually neomysin. (In fact, disrupting the functional gene (i.e., mutating the gene of interest) is often accomplished by interrupting the gene's sequence with the antibody resistance gene.) If all cells are then grown on a medium containing the antibiotic, those with no resistance (no insertion) will die. The second gene permits separation of those ES cells undergoing non-specific recombination from those undergoing homologous recombination. For example, the gene for thymadine kinase, *tk*, can be inserted into the vector outside the gene of interest. If non-specific recombination occurs, *tk* will be inserted in addition to the gene of interest. (Remember, if homologous recombination occurs, only the gene of interest is inserted.) By treating the remaining cells with a drug that is lethal to cells containing *tk*, we eliminate all but those in which the mutated gene of interest has replaced its wild-type homologue.

Finally! We now have *brown* mouse ES cells that contain the disrupted gene. However, they are heterozygotes. The gene on one chromosome is disrupted, but the other remains wild-type. The next step toward generating mice that are homologous for the disrupted gene is to inject these ES cells into *black* mice embryos. The embryos are transplanted into black surrogate mothers. The resulting offspring are referred to as "chimera." Because the embryos contain cells from both a brown and a black mouse, the offspring have a brown and black coat. Assuming that some of the original ES cells became germ-line cells, basic Mendelian genetics dictates the remaining steps. The mature male chimera is mated to a black female. Brown is the dominant coat color, and some of the male's gametes will have derived from original black cells, some from brown ES cells (with the disrupted gene). Black offspring—double recessives—obviously do not have the gene of interest. But only some of the brown offspring do. (Remember that the original ES cells were heterozygous for the gene of interest.) Simple genetic analysis reveals which of the brown mice carry the gene of interest, but even these mice will be heterozygous, as one copy of the wild-type came from the black mother to which the chimera was mated. A final round of crossing heterozygotes will generate a proportion of offspring who are true "knock-outs." These animals have two copies of the disrupted gene of interest.

Researchers constructing the first knock-outs confronted an immediate and difficult challenge. Consider the general research objective: to understand a gene product's function in a particular tissue at a particular time. But the first knock-outs were in essence genetic lesions, and familiar methodological objections to lesion studies applied to them. A single gene (and its protein product) can function in many pathways or processes in different tissues at different times over the course of an animal's development. To conclude that the disrupted gene causes some malady that a knock-out suffers can fail to acknowledge the gene's "downstream" effects. Furthermore, disrupting some genes during development is lethal. Genetic engineers solved these problems by developing "conditional" knock-outs. Using more sophisticated versions of the plasmid vector techniques, molecular biologists now construct vectors inserting mutated genes that can be turned on or off in only specific tissues and/or at precise times. Typically the trigger is a pharmacological agent, but in the next chapter we'll see another mechanism employed (a promoter turned on by heat shock).

We turn next to transgenics. The fundamental difference between knock-outs and transgenics is the specificity of the gene insertion. In knock-outs, a wild-type gene is replaced by a mutated, nonfunctional version through homologous recombination. This specific replacement occurs at a very low frequency. The frequency of random DNA insertions is much higher. In fact, many copies of a transgene can be inserted into a single cell's genome. Organisms whose cells contain these many insertions—up to 100 copies per cell—are referred to as "transgenics." Here the gene of interest contained within the plasmid vector is injected into the two nuclei of a fertilized egg (one from each parent) before they fuse. Because development is at the single-cell stage and the transgene insertion is non-specific, there is no need to select for specific insertion (as there was with knock-outs). Once fused, these fertilized eggs are injected into a pseudopregnant female (which has received hormone treatments to make her receptive to the eggs). The eggs divide and differentiate normally. 10-30% of the offspring will contain the inserted foreign DNA (the transgene) in all cells of all tissues, including germ tissues. Each cell will contain equal numbers of the gene insertion. Non-specific insertions can disrupt normal development and function, but typically 10-20% of these mice will breed normally. Using breeding schemes similar to knock-out technology, mice can be crossed and backcrossed to produce pure transgenic lines. In transgenics, the protein product of the inserted gene is overexpressed compared to wild-type controls. This expression is in addition to whatever products are expressed by other genes on the chromosomes on which the trangene was inserted.

One cannot overemphasize the impact of biotechnology across all of contemporary biology. The last quarter-century has revolutionized the discipline. This impact has resonated through neuroscience. As Karl Herrup has remarked, "rapid and complementary advances in the fields of molecular biology and experimental embryology have combined to offer neuroscience researchers unprecedented power to manipulate the mammalian genome"

(Zigmond *et al.* 1999, 421)—and with it, the protein products that drive neuronal activity and plasticity. This subsection is only intended to give a sense of the techniques that have come to dominate mainstream neuroscientific research. In the next subsection, we'll see in specific detail how these techniques have been employed to engineer animals that tie the molecular mechanisms of L-LTP sketched above directly and experimentally to memory consolidation in behaving animals.

5.3 An experimental link between molecules and behavior: PKA, CREB, and declarative long-term memory consolidation

Armed with biotechnology's ability to engineer knock-out and transgenic mice, the long-sought experimental link between the purported molecular mechanisms of L-LTP and declarative memory consolidation appears to have been forged. First, Rusiko Bourtchouladze and Alcino Silva's group at Cold Spring Harbor generated mice in which the gene expressing CREB-1 was partially knocked out (Bourtchouladze *et al.*, 1994). Then Eric Kandel's group at Columbia University generated transgenic mice partially expressing a cloned gene transcribing a protein that keeps catalytic subunits of PKA bound to regulatory subunits (Abel *et al.*, 1997). Using an ingenious behavioral paradigm that simultaneously trains mice on a hippocampal-requiring (hence declarative) and a non-hippocampal-requiring (hence nondeclarative) task, both groups demonstrated intact E-LTP in hippocampal neurons and short-term memory on the declarative task but deficit L-LTP and long-term memory in the engineered mutants. Given the details of these studies, especially the careful controls both groups employed, it is difficult to resist the conclusion that "the PKA pathway is critically important for the *consolidation* of short-term *memory* into protein synthesis dependent long-term *memory*, perhaps because PKA induces the transcription of genes encoding proteins required for long-lasting synaptic potentiation" (Abel *et al.* 1997. 615; my emphases).

The CREB^{-} mouse mutant results from standard homologous recombination in embryonic stem (ES) cells (described in the previous subsection). The targeting vector consisted of cloned mouse CREB genomic DNA and a promoterless neomycin-resistant (*Neo*) gene. The *Neo* gene is inserted into an exon common to all known CREB mRNA isoforms. Chimeric males are mated with wild-type females to produce ES cell-derived offspring. Heterozygotes for the gene target were mated to generate homozygous CREB^{-} mutant mice. Standard molecular biological analyses revealed no CREB expression in CREB⁻ mutants. Nevertheless, mutants surviving to adulthood displayed a normal phenotype and behavioral repertoire, no histological or morphological defects, and no impairments in growth or development (Hummler *et al.* 1994).[26]

Informing the background of Bourtcholadze *et al.*'s (1994) studies with the CREB⁻ mutants were

- results with nonspecific protein and RNA inhibitors that effectively blocked long-term memory while leaving short-term memory intact;
- results with cultured neurons from the invertebrate *Aplysia* (a sea slug) indicating that CREB antagonists bound to CRE sites selectively blocked long-term synaptic facilitation while leaving short-term facilitation intact (reviewed in the next chapter);
- results with *Drosophila* (fruit fly) mutants indicating a role for each step in the cAMP-PKA-CREB pathway in odor conditioning long-term memory (also reviewed in the next chapter).

Could CREB be implicated behaviorally in mammalian memory consolidation? CREB⁻ mutant mice seemed a promising experimental preparation.

Bourtcholadze *et al.* (1994) report that their CREB⁻ mutants show no gross hippocampal anatomical defects or abnormalities when coronal brain sections are matched with wild-type controls. They also report that the mutants matched controls' performance on tests for foot shock sensitivity. In one-trial-per-day training for fifteen days in the Morris water maze, CREB⁻ mutants were significantly impaired on time to find the submerged platform, percent time in the correct quadrant during a probe task after the fifteenth training day when the platform was removed (the quadrant of the pool where the platform had been on every previous training session), and number of times crossing the previous platform location during the probe task..[27] However, most declarative memory tasks, including the Morris water maze, take several days for rodents to learn. Their temporal dimensions make it difficult to track and compare short-term memory with long-term memory, and hence the mechanisms of memory consolidation, experimentally in behaving animals. What is needed to test the specific effects of the CREB⁻ knockout on mammalian declarative memory consolidation is a hippocampal-dependent task that mice can learn quickly and retain for a period that exceeds short-term memory.

Bourtcholadze *et al.* (1994) came up with such a task. They subjected CREB⁻ mutant and wild-type control mice to a novel environment for two minutes; this "context" was a conditioning box with metal and Plexiglas sides, equipped with a removable electric floor grid. After the two minutes, a tone was sounded through a wall speaker for thirty seconds; this conditioned stimulus (CS) was at 85dB, well above the background noise in the chamber. After the tone ended, a foot shock was delivered for two seconds through the electric floor grid; this unconditioned stimulus (US) was .75 milliamperes, far above the sensitivity threshold for mice. Hence simultaneously and on a single trial, mice were learning an aversive contextual conditioning task—novel environment paired with foot shock US—and a fear conditioning task—tone CS paired with foot shock US. It has long been established that contextual conditioning in mammals requires an intact hippocampus, while fear conditioning does not. In these studies, the behavioral measure is percentage of time spent freezing for five minutes after being put back into the conditioning chamber (contextual conditioning) or hearing the tone in a different context (fear conditioning). Freezing is a stereotypical rodent fear response in which the animal crouches and ceases all externally observable movements (except breathing).

CREB$^-$ mutants and wild-type controls displayed statistically identical freezing responses immediately after US presentation. This indicates that immediate learning was not compromised by the lack of CREB in either hippocampus or amygdala. CREB$^-$ knockouts were also statistically similar to controls in freezing to both the context and the CS (tested separately) 30 minutes after initial training. Both groups froze on average around 40% of the five minutes to both the novel environment and the tone. This indicated that both declarative and nondeclarative short-term memory remained intact in the CREB$^-$ mutants. However, at both one hour and twenty-four hours after training, the CREB$^-$ mutants averaged only about 10% freezing time to the contextual cue (novel environment), while the wild-type controls still averaged around 40%. A similar result obtained for the fear conditioning experiment. At two hours and twenty-four hours after training, CREB$^-$ mutants showed equally less freezing than controls. The key to these results is that the loss of CREB function disrupts long-term memory for both contextual (declarative) and cued (nondeclarative) conditioning without affecting short-term memory. This is directly in keeping with the model of L-LTP induction sketched above and the hypothesis that these are the molecular mechanisms of memory consolidation in behaving animals.

Bourtcholadze *et al.* (1994) strengthened this argument by also studying the physiological responses of CREB$^-$ mutant hippocampal slices. Slices were prepared, kept alive, and measured for baseline response activity by standard methods sketched in the previous section. They induced LTP by stimulating Schaffer collateral fibers with a train of 100 electric pulses at 100 Hz and recorded field EPSPs from CA1 neurons. Wild-type control slices showed L-LTP response two hours past the inducing pulse train, while CREB$^-$ mutant slices had already decayed to baseline levels ninety minutes past LTP induction. Although initial response (E-LTP) was similar in CREB$^-$ mutants and wild-type controls, statistical differences in potentiated field EPSPs began appearing only ten minutes after the inducing stimulus. This is during the time that E-LTP processes are still occurring, directly in keeping with the current model of L-LTP induction sketched in the previous section. These are powerful experimental results, warranting Bourtcholadze *et al.*'s conclusion that "the behavioral and electrophysiological findings reported in this manuscript identify CREB as an important component of *memory consolidation* in mammals" (1994, 66; my emphasis).

Nevertheless, questions still remained. Bourtcholadze *et al.* (1994) did find a fine-grained deficit in E-LTP induction in the CREB⁻ knockouts following a single stimulus train. Also, since CREB is a ubiquitous transcriptional factor that is activated by second messenger pathways other than cAMP-PKA, the above results don't define a conclusive role in memory consolidation for the earlier steps in the L-LTP pathway. Bouercholadze *et al.* note these limitations explicitly when they conclude that "although in the CREB⁻ mutants, both LTP and memory decay within the first hour, it is unclear whether the decay in LTP is related to the decay in memory or whether these two phenomena are unrelated consequences of the CREB⁻ mutation" (1994, 64-65). Interestingly, engineered mice with ES knockouts of individual PKA subunit isoforms did not yield compelling data linking PKA to either LTP or long-term memory (Qi *et al.* 1996). There also remained the numerous methodological problems with research using knockouts beyond developmental stages. Neither Hummler *et al.* (1994) nor Bourtcholadze *et al.* (1994) found developmental abnormalities in the CREB⁻ knockouts, though of course this does not mean that subtle ones weren't present. There is also the complete nonspecificity of the CREB⁻ deletion; it was equally expressed in amygdala as well as hippocampus, and so could not be used to focus on the exclusive mechanisms of *declarative* long-term memory.

For all these reasons, Kandel's group adopted a transgenic approach (Abel *et al.* 1997). Transgenic mice were engineered by the pronuclear injection method (described in the subsection above) with a calcium-calmodulin kinase IIα (*CaMKIIα)-R(AB)* transgene. The *CaMKIIα* promoter, containing upstream control regions and the transcriptional initiation site, drives expression only postnatally and limits expression to forebrain sites, including hippocampus, neocortex, amygdala, and corpus striatum. The cloned *R(AB)* DNA (*cR(AB)*) codes for an inhibitory form of the Riα subunit of the PKA molecule. PKA molecules composed of this inhibitory subunit carry mutations in cAMP binding sites and act as a dominant inhibitor of both types of catalytic subunits. These PKA molecules thus resist cAMP-driven release of catalytic PKA subunits and block further steps in the cAMP-PKA-CREB pathway. Founder animals receiving the pronuclear injection bred successfully and standard biochemical analyses revealed that the offspring generations (*R(AB)-1*, *R(AB)-2*) received and expressed the transgene.

Abel *et al.* (1994) used a standard *in situ* hybridization technique to reveal the regions of *R(AB)* transgene expression throughout the brain. This technique utilizes the "recipe" for the (mutated) protein coded as a sequence of nucleotide bases that make up the transcribed mRNA. Since this sequence is known for R(AB) (after all, the transgene had been cloned), a piece of radioactive mRNA containing the complementary nucleotide sequence could be synthesized. Slices of brain tissue from various regions are exposed to the radioactive RNA, which binds to the molecules of transcribed mRNA from the expressed transgene. The slices are then mounted on slides and exposed to a solution containing a radioactive antibody for the protein. The slides are then rinsed, leaving molecules of radioactive RNA in cells that contain the complementary mRNA. These slides are developed using standard photographic emulsion. The radioactivity exposes the emulsion, and regions of the brain where the transgene was expressed appear as bright images on photographic plates. The last part of this process is called *autoradiography* and is standard fare in molecular genetics research. These techniques, using a radioactive oligonucleotide that binds to base pair sequences present in an untranslated leader within the transgene, revealed *R(AB)* expression throughout the forebrain of the engineered mice, especially in the hippocampus and neocortex (Abel *et al.* 1997, Figure 1A, B, C).

The PKA transgenics were next tested for PKA activity in the hippocampus and for L-LTP in Schaffer collateral-CA1 slices. Both in the presence and absence of cAMP, PKA activity in transgenic hippocampus was reduced by more than 25%, compared to littermate wild-type control (Abel *et al.* 1997, Figure 2). L-LTP in hippocampal slices, measured in the standard way via field EPSPs to Schaffer collateral stimulation for up to three hours following pulse-train induction stimulation, revealed that activity fell to pre-stimulation baseline in R(AB) transgenics in less than two hours. L-LTP in slices from wild-type littermate controls remained heightened and constant for the entire duration of the study (three hours after inducing stimulation) (Abel *et al.* 1997, Figure 4). Intact transgenics were deficient in the Morris water maze, using measures and probe trials similar to Bourtcholadze *et al.*'s (1994) study. These results are consistent with PKA's role in the cAMP-PKA-CREB pathway in L-LTP induction and maintenance, as well as L-LTP's hypothesized role in declarative (hippocampal-requiring) long-term memory.

But it was on the contextual-cued conditioning task borrowed from Bourtcholadze *et al.* (1994) that the most convincing data with the R(AB) transgenics emerged. Abel *et al.* (1997) found that the transgenics were not deficient, compared to littermate wild-type controls, on either immediate learning or short-term memory on the contextual conditioning task. All groups spent the same amount of time freezing immediately after the initial foot shock and when placed back into the novel environment (training box) one hour later. However, when replaced back into the context twenty-four hours later, transgenics froze for only about one-third of the time that wild-type littermate controls did, indicating significant reduction in declarative (hippocampal-requiring) long-term memory. In the CS fear conditioning (nondeclarative) task, transgenics displayed normal short-term memory when tested to the tone CS 1 hour after initial training (compared to wild-type littermates), and displayed normal long-term memory to the tone when tested twenty-four hours later. That is, R(AB) transgenics spent the same percentage of time freezing after tone presentations as did wild-type littermate controls. Finally, transgenic performance statistically matched that of wild-type mice treated with anisomycin, a potent nonspecific protein synthesis inhibitor, delivered by injection either thirty minutes before or immediately after context and cued training (Abel *et al.* 1997, Figure 6). This match strongly suggests that the same molecular mechanisms are affected by the specific transgenic manipulations and the nonspecific pharmacological intervention.

One key to Abel *et al.*'s (1997) results is the dissociation between consolidation switches for declarative and nondeclarative long-term memory, the first being compromised while the second was preserved in the R(AB) transgenics. Presumably this result was due to a lower level of transgene expression in the amygdala. Not only does this further the case for the hypothesis that the cAMP-PKA-CREB molecular pathway is the consolidation switch for declarative long-term memory in behaving mammals, but it also removes possible confounds that the behavioral deficits are due to visual, motivational, or motor coordination difficulties. R(AB) transgenics were able to learn and retain the CS fear conditioned response, whose visual, motivational, and motor demands are similar to the contextual conditioning task.

Armed with these experimental results, and embedded within the model of L-LTP we sketched in the previous section, Abel *et al.* (1997) draw a bold conclusion:

> Our experiments define a role for PKA in L-LTP *and long-term memory*, and they provide a framework for *a molecular understanding of the consolidation of long-term explicit memory* in mice. ... The consolidation period is a critical period during which genes are induced that encode proteins essential for stable long-term memory. The long-term memory deficits in *R(AB)* transgenic mice demonstrate that PKA plays a role in the hippocampus in initiating the molecular events leading to the consolidation of short-term

> changes in neuronal activity into long-term memory. (1997, 623-624; my emphases)

The 2000 Nobel Prize committee agreed. In combination with his earlier (and continuing) groundbreaking work on the molecular basis of memory using the sea slug *Aplysia* (discussed in the next chapter), Eric Kandel was awarded a share of the 2000 Nobel Prize for Physiology or Medicine. In his presentation speech at the ceremony, Urban Ungerstedt remarked:

> I am convinced that you and I will remember this Nobel ceremony for many years. This is because of the dopamine which Arvid Carlsson discovered, enabling the brain to react to what we see and hear; the second messengers that Paul Greengard described, carrying the signals into the nerve cell; and the memory functions that Eric Kandel found to be due to changes in the very form and function of the synapses... Eric Kandel's work has shown us how these transmitters, through second transmitters and protein phosphorylation, create short- and long-term memory, forming the very basis for our ability to exist and interact meaningfully in our world. (http://www.nobel.se/medicine/laureates/2000/presentation-speech.html)

My mentioning this award is not a bald appeal to authority. I offer it only to indicate the importance that the scientific community has placed on Kandel's and his colleagues work. I should also point out that this work is not without controversy in the neuroscience community. Controversy remains about whether the account I described provides the correct molecular mechanisms *of LTP*, above and beyond the more familiar controversy about LTP's being the reductive link for memory consolidation (Lynch 2000). For example, Gary Lynch and his colleagues have stressed the importance of activity-dependent cell adhesion molecules in breaking down and then reconstructing the dendritic spine in an entirely local fashion. Their data are intriguing (e.g., Bahr *et al.* 1997, Staubli, Chun, and Lynch 1998). Nevertheless, no matter how such controversies pan out, all serious alternatives now on the table about the molecular mechanisms of LTP exhibit the same reductionism about the "LTP-memory consolidation switch" (McGaugh, 2000). "Ruthless reductionism" is a cornerstone of this parade case from recent molecular neuroscience.

I close this section with a final remark, before we leave the scientific details for a philosophical interlude. I remarked at the beginning of this chapter that many philosophers of mind and cognitive scientists bemoan "how little we know about how the brain works." To hear these folks talk, it is as though we're still flummoxed by the brain's complexity and waiting anxiously for some novel experimental breakthroughs or theoretical insights. I hope that by reaching this point in this book, readers now reject this common misunderstanding. Current "ruthlessly reductive" neuroscience knows *a lot* about "how the brain works," at least about its basic constituents and how they interact; and it possesses tools of discovery that we can confidently expect to increase our knowledge. That this knowledge and these tools have been ignored by higher level theorists about mind—even by some who fancy themselves "philosophers of neuroscience"—is an interesting datum for "sociologists of philosophy." Some might find it difficult in the abstract to see how mind reduces to molecules; but one will never get over that intellectual hurdle or show conclusively why such reductions can't obtain by remaining ignorant of the best existing scientific attempts to do exactly this.

6 THE NATURE OF "PSYCHONEURAL REDUCTION" AT WORK IN CURRENT MAINSTREAM (CELLULAR AND MOLECULAR) NEUROSCIENCE

Perhaps this case doesn't meet everyone's pre-theoretical notion of what "reduction" amounts to or requires. Philosophers might ask: Where are the cross-theoretic bridge laws? Where are the explanatory generalizations or laws that compose the two theories? Where is the logical derivation of reduced from reducing theory? I've already proposed that we set aside preconceived notions and consider what the relation amounts to in this detailed example of an actual scientific reduction-in-practice. To generalize from this example, I will employ the metascientific resource of explanatory posits becoming "structured through a reduction" to lower level sciences. In the "structuralist" tradition in the philosophy of science, C.U. Moulines (1984) has suggested such a resource, and in Bickle (2002) I develop it further by applying some quasi-formal machinery to illuminate an abbreviated version of the memory consolidation-molecular mechanisms of LTP link. Here I will ignore the formalism and elaborate on the conceptual details.

In the previous sections we've heard much of cross-theory psychological-to-cellular/molecular "links." Can we cash that metaphor? We can say at least this much: entities, properties, and events postulated in psychological explanations (as in higher level theories generally) typically get characterized *functionally*, purely in terms of their *causes and effects*, with little or no explicit concern for the underlying neurobiological (lower level) events and processes producing this functional profile. Memory consolidation is a process that typically requires multiple stimulus presentations or rehearsals and can be disrupted by certain events that occur during a short temporal interval after presentations (its causes). It transforms a labile, easily disrupted short-term memory into a stable, durable long-term memory with related memorial or cognitive content (its effects). In this fashion, the psychological concept resembles that of "gene expression" in Mendelian through transmission genetics. The latter process is there characterized in terms of trait ratios in various breeding combinations across parent and offspring populations using theoretical posits—'allele,' 'dominance'—defined in their terms. In current molecular genetics, however, these purely functional concepts get linked to elaborate sequences of molecular and biochemical transcriptional, translational, and recombinatory pathways. This linkage is analogous to the way psychology's "memory consolidation switch" posit gets linked to the intracellular molecular processes of E- to L-LTP, occurring in hundreds of thousands of selective neurons to alter synaptic efficacy and ultimately behavior. But these examples only set up a problem for metascientific analysis, and philosophers of science have not carried this idea of posits "structured through reduction" beyond a few illustrations. I propose to go further, using the details of the example provided in this chapter.

It can be useful to characterize scientific theories in terms of *models* and *intended empirical applications*. A theory's models are the systems, both real world and mathematical, that share the structure characterized by the theory's fundamental assumptions and explanatory generalizations. Its intended empirical applications are all those real-world systems to which at any given time the theory is thought to apply (by the appropriate scientific community), the systems thought to possess the structure of the theory's full contingent of fundamental assumptions and generalizations. At any given time, some of a theory's intended empirical applications will have been shown to be actual models, i.e., to have the structure characterizing the theory's models. Other intended empirical applications will not yet have been shown to be models. Some of the latter will be in the process of empirical test.[28]

Using this scheme, a theory's models can be usefully characterized as consisting of three components:

- "empirical" or "base sets," whose elements are grouped into the fundamental categories by which the theory "carves up" the world;
- "auxiliary sets," the mathematical or other abstract spaces the theory employs in its explanations; and
- fundamental relations and functions, defined as operations over elements of the other two components, characterizing the theory's basic explanatory properties and relations.

All these components, but especially the last type, combine to specify the theory's basic assumptions and generalizations. The structure of the latter characterizes the structure of the theory's models.

A simple example helps to illustrate these abstract descriptions. Consider the theory of classical collision mechanics. Each of its models are composed of two empirical base sets (a set of particles and a set of time instances), one auxiliary set (the set of real numbers), and two fundamental relations (the mass relation, which assigns positive real numbers to elements of the particle set; and the velocity relation, which assigns ordered triples of real numbers to particles at time instances). These elements combine into the theory's one law, the conservation of momentum before and after a particle collision (namely, the sum for all particles p of the mass of p times the velocity of p at time instance t_1 equals the sum for all particles p of the mass of p times the velocity of p at time instance t_2, where t_1 is prior to and t_2 is after a particle collision). Any system, real world or mathematical, that possesses this structure is a model of classical collision mechanics. Its intended empirical application is to every real world system composed of particles with mass and velocity and time instances. (Theories in the special sciences have less broad sets of intended empirical applications.)

From this perspective on the structure of theories, reduction is partly a mapping of specific models and intended empirical applications across the two theories that meets a variety of limiting conditions (not any old mapping constitutes a reduction relation). One of these limiting conditions is the requirement of *mappings of each empirical base set of the reduced theory into a base set, a fundamental relation, or combinations of the two of the reducing*, guided by the two theories' shared intended empirical applications.[29] In some theory reductions, empirical base sets of models of the reduced theory get mapped onto identical base sets of reduction-related models of the reducing (in the extensional, set-theoretic sense of identity—the mapped sets contain the same elements). For example, the set of planets in the intended empirical application of Kepler's astronomy is identical to the set of planets in the reduction-related intended empirical application of Newton's celestial mechanics. But typically, in interesting scientific reductions, some (if not all) of the empirical base sets of models of the reduced theory get mapped onto *nonidentical* sets making up the reduction-related models of the reducing. Examples from the history of science abound (Moulines 1984). In the reduction of rigid body to classical mechanics, the empirical base set of rigid bodies in models of the former gets mapped to that of Newtonian particles in reduction-related models of the latter. But elements of the former set are not elements of the latter. Neither set identity of inclusion holds across this link. No rigid body is itself an element of any set of Newtonian particles.

A second example is even more analogous to our case study, where a single empirical base set in models of a reduced theory gets linked to a sequence and combination of both base sets and fundamental relations in the reduction-related models of the reducing theory. This is the reduction of Mendelian to molecular genetics.[30] Elements of the base set of genes, characterized solely (i.e., purely functionally) by Mendel's laws of Segregation, Independent Assortment, and Dominance, are not members of any empirical base sets in models of molecular biology. The latter are sets of organic molecules and the theory's fundamental relations specify their interactions to generate DNA transcription, messenger RNA translation, protein synthesis, and phenotype production. The two ways that the reduced and reducing theories "carve up" the world of genetic inheritance and expression differ, as expressed in the different elements composing their models' empirical base sets, fundamental relations, and explanatory generalizations. Nevertheless, the base sets and fundamental relations making up models of these two theories get related as a condition on the global theory reduction relation across models of the two theories, and the two theories' intended empirical applications (the real-world biological systems expected to have the structure of models of the two theories) are virtually identical.

In cases like this, where an empirical base set in models of the reduced theory get linked to sequences and combinations of base sets and fundamental relations in reduction-related models of the reducing, we also get an account of what it is for "amorphous" basic entities of the reduced theory to become "structured through reduction." An entity or process, characterized entirely by way of the fundamental relations and generalizations of the reduced theory, comes to be related *in a domain eliminating way* to sequences of entities and processes characterized by the relations and generalizations of the reducing theory. From the perspective of their reducing theories, there are no rigid bodies, separate space and time, or Mendelian genes. Elements of these empirical base sets aren't elements of the empirical base sets or fundamental relations and functions by which their reducing theories (respectively, classical particle mechanics, special relativity theory, and molecular genetics) "carve up the world." However, the roles that elements of these empirical base sets play in the relations and generalizations that partly define models of the reduced theories bear interesting structural similarities to the roles played by sequences and combinations of empirical base sets and fundamental relations in reduction-related models and empirical applications of the reducing theory. In particular, entities characterized on the reduced theory *primarily by their functional (input-output) features* get linked to complex structures (sequences and combinations of entities and processes) whose dynamics and interactions specifiable entirely within the reducing theory

1. apply to roughly the same intended set of real-world systems, and
2. provide causal mechanisms that explain the former's functional (input-output) profile.

This, in rough outline, is the metascientific concept of a theoretical posit becoming "structured through reduction."

We can abstract this metascientific resource out of the detailed memory consolidation-molecular mechanisms of LTP reduction. Models of a psychological theory of memory, by their empirical base sets, fundamental relations, and generalizations characterized in terms of these, posit an entity/process, the consolidation switch. But they characterize this posit only in terms of the time course and amount of repetition needed to convert a given type of memory item from short-term memory to long-term memory and the behavioral efficacy of different types of retrograde interference. In other words, psychology characterizes this entity/process in purely functional fashion, in terms of the events causing it and the events it in turn causes, with no concern for the *specific underlying neural mechanisms* generating this functional profile. (Notice also that psychology does the same for short-term memory and long-term memory, characterizing both primarily in terms of duration and distractability of recall after initial presentation.) Cellular and molecular neuroscience then developed an account of the mechanisms of LTP, and behavioral neuroscience verified that these mechanisms underlie memory consolidation selectively in behaving animals (using, e.g., the Kandel lab's R(AB) transgenic mice). The consolidation switch empirical base set in models of psychology got mapped to sequences and combinations of empirical base sets and fundamental relations in reduction-related models and intended empirical applications of molecular neuroscience, in particular to those involved in the transition of E-LTP into L-LTP and the maintenance of L-LTP. The empirical base sets of the latter include intracellular and neural transmission molecules: adenylyl cyclase, cAMP, PKA, CREB enhancers and repressors, DNA, RNA polymerases, ubiquitin hydrolase, CCAAT enhancer binding protein, glutamate, dendritic spine cytoskeleton components, AMPA receptors, NMDA receptors, and so on. Out of these components, the fundamental relations of cellular/molecular neuroscience build membranes, plastic neurons, neural circuitries from sensory input to motor effectors, neuromuscular junctions, and the specific connections with the muscles and skeletal systems (themselves built up by fundamental relations out of their molecular elements). Molecular neurobiology's fundamental relations and generalizations also specify these elements' subcellular physiological and extracellular interactions: phosphorylation, selective molecular bindings, gene transcription and translation, protein synthesis, changed receptor affinity for ion conductance, changing protein configuration, ions conducting through selective protein channels, and neuromuscular interactions leading to combinations of muscle contractions pulling against a calcium frame (the skeleton).

The mapped posits of the two theories play structurally similar roles across reduction-related models and intended empirical applications, namely, mammals behaving in observable, quantifiably measurable ways in tasks involving stimulus exposures earlier in time. However, elements of psychology's "consolidation switch" empirical base set are not elements of the sequences and combinations of molecular neuroscience's empirical base sets and fundamental relations across reduction-related models and intended empirical applications. Cellular and molecular neuroscience "carves up the world" of behavioral memory data in a fundamentally different way than psychology, even though the intended empirical applications of the two theories are virtually identical. The detailed scientific example of reduction-in-practice described in this chapter gives us a beautiful example of a psychological posit—the consolidation switch—"structured through reduction" to a molecular process—the mechanisms of L-LTP—occurring in the neurons comprising the appropriate anatomical circuits. Such is the metascientific nature of the reductionism both endemic to and successful within current mainstream neuroscience.[31]

A useful visual metaphor of this abstract cross-theory link is two nets composed of strings tied onto rings (Figure 2.11). The rings are analogous to a theory's posits (empirical base sets and fundamental relations) while the connecting strings are analogous to the causal connections its generalizations postulate to hold between them. The rings of the reduced theory have significantly greater diameter than the rings of the reducing, and the lengths of strings between the reduced theory's rings are considerably longer. This reflects the much more coarse-grained, purely functional way that the reduced theory "carves up the world" (the shared intended empirical applications) compared to the reducing. Notice that the specific ring sizes and string lengths making up the reduced theory do not belong to the reducing theory. In the latter, there are no rings of those specific diameters or string segments with those specific lengths and connectivity patterns. But if we lay the reduced net on top of the reducing—the analog of relating an intended empirical application of the reduced to one of the reducing theory, as part of the theory reduction relation—we see which collection of reducing rings and strings (posits and causal connections between them) gets located underneath the reduced rings and strings. Applied to our case study, we see *how* (and not just *that*) the emerging cellular/molecular story arranges its constituents together into a sequential and combinatorial structure abstractly similar to psychology's coarse-grained functional posits. In our example, psychology's consolidation switch—one type of big ring—is an important functional *approximation* of the cellular/molecular mechanisms that signal the switch from E-LTP to L-LTP and maintain L-LTP selectively in many synapses throughout the neuronal network. Visual metaphors are crude and have their obvious limitations, but this one captures some important aspects of the nature of psychoneural reductionism-in-practice in our detailed example from current mainstream neuroscience.[32]

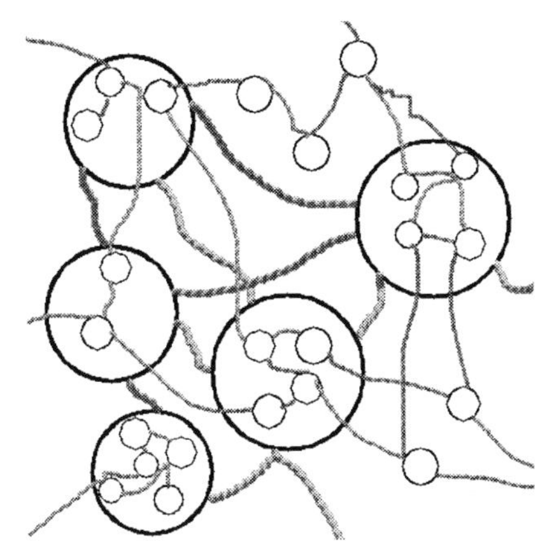

Figure 2.11. Visual metaphor for the structure of intertheoretic (scientific) reduction. Large rings represent posits of the reduced theory; the thicker strings connecting them represent posited causal relations by that theory's explanatory generalizations. Small rings represent posits of the reducing theory; thinner strings connecting them represent posited causal relations by that theory's explanatory generalizations. See text for detailed discussion. Figure created by Marica Bernstein.

If you are still not satisfied with the example of reductionism-in-practice presented in this chapter or the metascientific account I abstracted out of it, you might examine your reasons. Perhaps some prior epistemology you are applying to science leads you to a view about "what reduction has to be." Perhaps you are influenced by the nature of reduction in other disciplines (physics or mathematics). Perhaps you are guided to a theory of reduction based on prior ontological considerations about "what physicalism requires." Perhaps the resources you employ to theorize about science aren't neutral, but influence you to understand the reduction relation in a particular fashion, e.g., perhaps you still conceive of theory structure in terms of the syntax of predicate logic and so still think about intertheoretic relations in terms of first-order deductive validity. Or perhaps you're motivated by an external normative consideration; perhaps you want a justification, external to the results and accepted explanations of current science, e.g., of why this research is where funding should be going or what level of research neuroscientists should be pursuing.[33] Regardless of what might motivate dissatisfaction with my calling the scientific results reported in this chapter a "reduction," the facts revealed here are these. Psychoneural reductionism is alive and thriving in current cellular and molecular neuroscience, as revealed in the attitudes of its practitioners and their choices of experimental manipulations and investigations. And the specific nature of the reductionism encapsulated within these practices and results is usefully illustrated by the "structuring" of psychology's purely functional posits into specific sequences and combinations of cellular and molecular entities, processes, and causal interactions. That is reductionism-in-practice in current mainstream neuroscience and its metascience. And if it doesn't meet your borrowed ideal of "what reductionism has to be," perhaps first philosophy and *a priori* normative epistemology really IS your cup of tea. The target of this final comment is not just the "American Philosophical Association" crowd, but also many "naturalistic" philosophers of mind—despite the latters' denial of caring about such "traditional philosophical" concerns. With regard to declarative memory consolidation, "ruthlessly reductive" neuroscience is already down to the "molecular pathways" level.

NOTES

[1] It might seem strange to talk about the spinal cord and memory, but this has recently received some attention in the study of chronic peripheral pain. In fact, the same molecular mechanisms involved in cortical and hippocampal long-term potentiation underlie the experience-driven synaptic plasticity hypothesized to explain features of chronic pain. See Sufka (2001) for a clear exposition of this hypothesis.

[2] If these scientific details elude you now, fear not. Later in this chapter I'll explain in detail the molecular mechanisms of LTP and experimental protocols and results in studies of this sort. For novices, I'll even include a brief "primer" on basic cellular neuroscience.

[3] Schouten and de Jong inform me that this argument in their (1999) was intended to reveal the cross-categorizations that obtain between psychological and neurobiological concepts, rendering any sort of "bottom-up" explanatory approach untenable. If so, then their argument rests upon issues I will take on in Chapter Three, namely, questions about methodology within neuroscience dominated by the search for cellular and molecular mechanisms and the issue of multiple realization.

[4] Many readers will be familiar with well-known exceptions to the stimulus repetition feature. One-trial learning that remains stable for long periods has been studied for more than four decades. Typically, these are species-specific and evolutionarily prominent learning processes, e.g., conditioned taste aversions in rats. See, e.g., Garcia and Koelling (1966) (although most reputable learning theory textbooks will include a description of this phenomenon). For ease of exposition, I won't be qualifying the "stimulus repetition" condition on consolidation in the discussion to follow. Fans of one trial learning can make the appropriate mental adjustments to my assertions, e.g., that stimulus repetition and rehearsal typically improves long-term memory

recall and performance. Thanks to Carl Craver for pointing out the need to qualify some remarks I make in the discussion below.

[5] This study is summarized nicely in Squire and Kandel 1999, 130-132.

[6] Duncan also measured the latency of time spent in the test compartment after rats were placed there initially, but got little useful information from this behavioral measure and didn't include it in his published results.

[7] This point is not the only thing that "cognitivists" and "autonomists" could and do say. We'll return to this point in earnest in the first two sections of Chapter Three. This discussion also foreshadows that about declarative memory, later in this chapter.

[8] I'll explain some of these biotechnological manipulations later in this chapter.

[9] Cajal's original publication is in French. Squire and Kandel (1999, 35-36) give a brief report of Cajal's speculations.

[10] Material presented in this subsection is abbreviated standard neurobiology textbook fare. Any reputable recent textbook will provide additional details. For those interested in a state-of-the art account of how neurons work, I recommend Levitan and Kaczmarek (2001).

[11] The other type is neuroglia, a kind of connective tissue. The many roles of glial cells, and neuron-glial interactions, are targets of much current research. Much of this is beyond the scope of our concerns.

[12] The "spike" metaphor describes the shape of the event recorded on an oscilloscope, as pictured in Figure 2.3.

[13] Obviously, the "presynaptic terminal" of the primary synapse is postsynaptic to the modulatory neuron. More recently discovered complexities render the original "pre-" and "postsynaptic" terminology confusing.

[14] Thanks to Carl Craver and Ken Sufka for reminding me of this point. Sufka points out in particular the new evidence of "pattern coding" in axons above and beyond rate/frequency coding of their action potentials (e.g., in Reichling and Levine 1999). I have not studied this literature carefully; nevertheless it is difficult for me to see how it escapes the general point I'm urging in this paragraph (and the next). At bottom, however, this is an empirical issue. If someone can show me specifically where or how this argument is wrong in light of some new, empirically confirmed coding discovery, I'll give up this version.

[15] This subsection is not intended to be a comprehensive early history of LTP. Philosopher of neuroscience Carl Craver has a forthcoming manuscript (2003) that explores this history in excellent detail. Part of his history is based on interviews with key participants.

[16] Notice that this is amplitude increase in *population spike*, not in individual spikes. (Recall the all-or-none law of the individual action potential.) This means that the single stimulating pulse to the perforant path caused more dentate neurons to reach threshold of excitation after the stimulus conditioning train had been delivered, compared with before.

[17] I'll say more about the hippocampus and declarative memory in the next section, with references to the recent literature.

[18] The molecular genetic details are not crucial here, but will become so later in this section. I will then provide a brief primer on the molecular mechanisms of gene transcription.

[19] A receptor *antagonist* blocks the transmitter's effect, typically by competing with transmitter molecules for receptor binding sites but not initiating the transmitter's activity when bound. On the other hand, a receptor *agonist* increases the efficiency of transmitter molecules at the receptor site.

[20] This much molecular genetics is sufficient for our purposes in this section, but I'll present more details in section 5.2 of this chapter when I turn to biotechnology's contribution to the neurobiology of memory consolidation. For readers interested in pursuing this fascinating area more thoroughly, Lewin (1999) is a state-of-the-art textbook for current molecular genetics, but any reputable introduction to biology text will contain a discussion of molecular genetics in plenty of detail for the average non-scientist. Consult your local Biology department.

[21] CREB proteins only have their transcriptional effects when phosphorylated. Recall from above that catalytic PKA subunits have translocated to the neuron's nucleus. These translocated subunits phosphorylate CREB proteins.

[22] Clayton (2000) is a useful recent review, although his "operational definition" of 'immediate early gene' does not square exactly with its usage by other molecular neuroscientists.

[23] A detailed discussion of celebrated case H.M. with references to the primary literature can be found in any reputable neuropsychology textbook, e.g., Kolb and Whishaw (1996), 357-360.

[24] Subsection 5.2 is co-authored by Marica Bernstein.

[25] Recall from the brief description in subsection 4.2 of this chapter the basic dogma of molecular genetics: DNA→(transcription)→RNA→(translation)→protein.

[26] This raises an interesting question, especially given CREB's ubiquitous occurrence in all types of biological tissues. Why aren't the knockouts more deficient? Hummler *et al.* (1994) argue that CREB operates in tandem with two other cAMP-driven transcriptional activators, cAMP response element modulation protein (CREM) and activating transcription factor 1 (ATF 1), which can compensate for each other. CREB$^-$ mutants do in fact overexpress CREM in all tissue types.

[27] More intensive training—three blocks of four trials per day over three training days—overcame these statistical differences between CREB$^-$ mutants and controls. Bourtcholadze *et al.* (1994) suggest that this is due to CREM compensation for the CREB deficiency (mentioned in the previous footnote).

[28] This description constitutes an informal account of the structuralist program's concept of a theory-element. Most structuralists find that set theory and category theory provide useful mathematical resources to characterize these concepts formally. See Balzer *et al.*, (1987), Balzer and Moulines (1996), and Bickle (1998) for detailed presentations, including models of intertheoretic reduction (Balzer *et* al. 1987, chapter 5, Bickle 1998, chapter 3) and an application to earlier developments in the reduction of psychological theories of memory to neurobiological counterparts (Bickle 1998, chapter 6).

[29] The relations making up this condition are called "ontological reduction links" by Moulines 1984. I adopt this terminology in Bickle 1998, chapter 3, and 2002).

[30] That this is a case of reduction is very controversial in philosophy of biology. However, the molecular geneticists I know have no qualms about calling and treating it as such. Perhaps a reduction-in-practice of recent work in that discipline would be a useful contribution to philosophy of science.

[31] To see these features developed in quasi-formal fashion, see Bickle (2002).

[32] I've used a "net" metaphor, across changing concerns and philosophy of science backgrounds, to express ideas about theory reduction in psychology and neuroscience since one of my earliest papers (Bickle 1992b). See elaborations in Bickle (1998, chapter six).

[33] Thanks to Carl Craver for stressing this "normativity" worry.

CHAPTER THREE
MENTAL CAUSATION, COGNITIVE NEUROSCIENCE, AND MULTIPLE REALIZATION

We next trace implications of the detailed example from Chapter Two for two prominent issues in current philosophy of mind and one increasingly prominent area of current neuroscience. Reduction is central to all three. The first philosophical issue—the problem of mental causation—questions whether or how mental properties can exert causal effects on behavior. The second philosophical issue—the multiple realization argument—is widely thought to be one of two decisive arguments against the reduction of mind to brain or psychology to neuroscience.[1] We saw in the previous chapter that "reductionism" is alive and well in current cellular and molecular neuroscience. Now we'll see whether that reduction-in-practice and its results carry helpful implications for issues that have attracted serious philosophical attention. The scientific issue concerns the status of cognitive neuroscience vis-à-vis the discipline's cellular and molecular core. How do studies involving, e.g., neuron population dynamics or specific activations across neural regions, relate to ones exemplified by our detailed example from the previous chapter?

1 THE PROBLEM OF MENTAL CAUSATION

"Mental causation" is an ideal problem for modern-day philosophers of mind and cognitive science. That our mental states are causally efficacious for our behavior is central to our shared self-conception. Ruth answered "B" to question #3 on the personality assessment *because* she *believes* that she is kindly. Alonzo signaled the waiter *because* he *wants* another martini. The simplest reading of these common assertions is that the mental properties of our cognitive states—e.g., the contents of Ruth's belief and Alonzo's desire—were part of the nexus that generated the action (including the limb movements). Had Ruth's or Alonzo's causally efficacious belief or desire contents been different, different behaviors would have resulted; perhaps arm movements with the pencil that result in Ruth's drawing a circle around answer "C," and ones that result in Alonzo's brining the martini glass up to his open mouth. Yet this common feature of our self-conception raises delicious puzzles. For example, are mental contents *in their capacity as* (or "*qua*") mental contents (and not, say, in their capacity as the brain states that realize them) causally efficacious, or are content properties like the color of the benzodiazepine tablet with regard to the pill's tranquilizing effects? The soprano hits a high C note while uttering the English word "Shatter!" The glass shatters. Clearly the semantic properties of her utterance had no causal effects on the glass's shattering. That was entirely an effect of the event's acoustic properties on the microstructure of the glass's molecules. Had the event's semantic properties differed while the acoustic properties remained constant, the glass still would have shattered. Might mental properties be like that: causally *in*efficacious, with all the causal work on behavior being done by the event's physical (e.g., neuronal) properties? That suggestion is contrary to our self-conception of being the possessors of causally efficacious mental states. But *how* are mental properties different in a way that permits us to locate them justifiably in the causal fray? A little reflection on this puzzle and philosophers are off and running. The mental causation literature contains theories of "supervenient causation," "causal-explanatory exclusion," "downward causation," "ceteris paribus laws," "causal compatibilism," and other philosophical exotica. And from such humble origins![2]

To my lights, Terence Horgan (2001) has recently provided the best formulation of the metaphysical and epistemological intuitions driving both sides of the mental causation debate. He presents them as an *inconsistent quintet* of claims, each plausible individually but inconsistent when conjoined:[3]

1. Physics is causally closed. Each physical event is determined (to whatever extent it is determined) completely by purely physical events.
2. Mental properties are causal properties
3. Mental properties are not identical to physical causal properties, because the former are multiply realizable on the latter.[4]
4. Mental properties are real and instantiated in humans.
5. If physics is causally closed, then all causal properties are physical causal properties.

(By 'physical property' Horgan means the kind of property postulated in fundamental physics.) Any four of these intuitively plausible claims is consistent, but those four conjointly entail the falsity of the other. The "problem of mental causation" becomes which claim to reject. "Causal emergentists" deny claim 1, "epiphenomenalists" deny claim 2, "identity physicalists" deny claim 3, "eliminativists" deny claim 4, and "causal compatibilists" deny claim 5. Each view "solves" the problem of mental causation, but at the expense of an intuitively plausible metaphysical or epistemological claim.[5]

Some physicalist philosophers—reductive physicalists, mind-brain identity theorists, "central state" materialists, "brain state theorists"—are initially attracted to Jaegwon Kim's "causal-explanatory exclusion" arguments (1993, 1996, 1998). Using Horgan's (2001) schema, Kim's arguments defend claim 5. Physicalists so moved will thus opt for rejecting either claim 3 or 4. Horgan (2001) gives an insightful presentation of Kim's reasoning. First, since philosophical orthodoxy insists at least that all mental causal properties "supervene upon" or "are realized by" physical causal properties, the ultimate causes of any mental property's becoming instantiated are themselves physical properties. In particular, they will be whichever physical properties cause that mental property's physical realizers to occur. Second, if mental properties are both causal properties and distinct from physical properties, then either

A) some events depend causally (at least in part) on the prior occurrence of some mental property(ies),

or

B) some events are causally overdetermined by individually sufficient physical and mental properties.

If (A), then the effect-states can't be physical or else the causal closure of physics is violated. There would then be physical events for which physical causes would not be sufficient. But the effect-states can't be mental, either, without violating the demand of physical supervenience or realization of mental properties. There would then be mental properties for which physics does not provide a complete supervenient base or realization. So (A) is incoherent (for a "minimal" physicalist). Given the causal closure of physics, mental properties can only be causal properties via the physical properties that realize them. But that rules out (B) since there couldn't then be causal overdetermination in any real sense. There would then be no independent causal route leading from mental cause to effect. Kim's upshot is that the physical properties "do all the causal work." Physical causal explanations of behavior "screen off" or "exclude" mental causal explanations.

Unfortunately for physicalists, and despite its seeming reasonableness, Kim's causal-explanatory exclusion argument has been attacked relentlessly by the philosophy of mind orthodoxy. Tyler Burge (1993) and

Lynne Rudder Baker (1993) have urged that our commitment to mental causation, as reflected in our ordinary mentalistic explanatory practices, is far stronger than our commitment to abstract physicalist metaphysical principles. We should focus on the former and learn to love explicit mental causation. At least since his (1993) essay, Terence Horgan has argued that treating causal claims as reflecting "counterfactual dependencies" renders mental causal claims seemingly obviously true, despite any physicalist metaphysical scruples we might hold. We all readily agree that had Alonzo not desired a beer (counterfactually), then *ceteris paribus* he wouldn't be reaching right now for the Schlitz can in his fridge. Numerous philosophers have urged "the generalization problem" on Kim. Doesn't causal-explanatory exclusion do away with all causation "higher than" basic physics, including the neurobiological, molecular biological, and biochemical; and isn't a metaphysics that eschews that much simply reduced to absurdity? Physicalists of course have some retorts. Kim himself offers a number in chapter 3 of his (1998) book. But the debate quickly takes on the "fruitless clash of intuitions" flavor I vowed to avoid in Chapter One. I for one am not an enthusiast for wading into this argument here.

So I propose a different tack. I want to appeal to our detailed scientific example from Chapter Two and remind readers of the two distinct theoretical/explanatory levels at work: memory consolidation from experimental psychology and the molecular mechanisms of LTP from cellular and molecular neuroscience, both applied to behavioral data like that from the Kandel lab. I contend that when we fix our gaze on aspects of scientific practice in this actual recent example, we see that psychological explanations *lose their initial status as causally-mechanistically explanatory* vis-à-vis *an accomplished* (and not just an anticipated) cellular/molecular explanation. All attempts by philosophers to "save" mental causation presuppose that psychological explanations remain (causally) explanatory, at least in certain contexts, even in light of accomplished "low level" neurobiological explanations of the same behavioral data. I'll argue instead that within scientific practice, psychological explanations *become otiose* when the type of cellular/molecular explanation encapsulated in the detailed example and now dominant in mainstream neuroscience is achieved. There is no need to evoke psychological causal explanations, and in fact scientists stop evoking and developing them, once *real neurobiological* explanations are on offer (but not merely "on promise"). Philosophers who deny this point are usually guided by outdated accounts of real neuroscientific practice. Contra Kim, lower level explanations need not "exclude" higher-level accounts in any deep epistemological or metaphysical sense. But the former do *render the latter pointless, along with any further search for empirically improved successors at the same level*—except for some residual, purely heuristic tasks. I aim to show that accomplished lower-level mechanistic explanations absolve us of the need in science to talk causally or investigate further at higher levels, at least in any robust "autonomous" sense.

2 LETTING SCIENTIFIC PRACTICE BE OUR GUIDE

To articulate and defend these claims, I propose that we let scientific practice be our guide. As cellular/molecular accounts of memory consolidation emerge, like the Kandel group's, what becomes of psychological-level research on the topic? I'm asking here about the empirical search for *causal explanations* of behavioral data. The answer is, it disappears. Circa 2002, there is no "psychology of memory consolidation." No one is searching for empirical refinements or supplements of, e.g., Duncan's (1949) causal-explanatory resources (Chapter Two, section 2 above). Psychological research into the mechanisms of memory consolidation hit an explanatory wall throughout the middle of the last century and neurobiology (including molecular genetics) quickly took over.[6]

We must realize how surprising this shift should be for anyone who advocates the "autonomy" of psychology from neuroscience. Neuroscientists, and even many experimental psychologists, may be surprised to learn just how popular "psychological autonomy" is among philosophers of mind. It is the orthodox view at present, due to its connection with the dominant "nonreductive physicalist" solution to the mind-body problem. Memory consolidation would seem like a natural *explanadum* for the psychological autonomist. We saw in the previous chapter (end of section 2) how its basic empirical features—the typical need for stimulus rehearsal and the impact of retrograde interference—can be given natural "cognitive" interpretations. Rehearsal amounts to "calling to mind" repeatedly the representational content of stimulus presentations available in short-term memory. Retrograde interference disrupts the required maintenance of the contents of memory representations. Such accounts fit with the widely accepted "mark of the cognitive," as requiring explanations that advert to operations over the contents of cognitive representations. Furthermore, these principal features of memory consolidation are operative especially in mammalian "declarative memory," the "flexible" form compromised in human amnesia and its best current primate models (Chapter Two, section 5.1 above). In memory consolidation, especially as it occurs in the higher forms of memory prevalent in mammals, we appear to have clear examples of mental causation. The content properties of the short-term memories are part of the causal processes, primarily rehearsal, that generate long-term memories with specific, related contents. Investigating the nature of the "consolidation switch" seems surely a task for psychology, not neuroscience, and especially not low-level cellular and molecular neuroscience that investigates receptor biochemistry, membrane ionic conductance, and intra-neuron signaling pathways. Yet as we saw in the previous chapter, in great detail, the scientific facts are otherwise.

The lesson I drew in the previous chapter from the detailed scientific example pertained to the nature of real reductionism in real neuroscience. Now I draw a second lesson. In light of this existing cellular/molecular explanation and these experimental results, it seems silly to count psychology's "explanation" of consolidation as "causally explanatory," "mechanistic," or a viable part of any *current* scientific investigation *still worth pursuing*. The explanation, "Kurt remembered that telephone number today that I relayed to him yesterday because he rehearsed it mentally fifteen times without retrograde interference for thirty minutes after he heard it," pales in comparison to one that appeals to activity-dependent molecular (including molecular genetic) mechanisms occurring in millions of selective neurons during the "consolidation phase." In any case, that is the lesson from scientific practice. Recall from Chapter Two that what psychologists called "stimulus rehearsal" (and "repetition") amounts to increased activity and glutamate release in selective presynaptic axons, enhancing the amount of freed PKA catalytic subunits in the postsynaptic cell, permitting these subunits to translocate to the neuron's membrane and exert their molecular genetic effects that lead to long-term structural changes that lastingly potentiate the neuron's response to similar inputs later. What psychology called "retrograde interference" gets explained as processes initiated after initial stimulus presentations that interfere with any of the cellular/molecular steps generating L-LTP. The Kandel lab's ingenious behavioral data, using PKA regulatory subunit transgenic mice as the key experimental group, establishes that this molecular pathway is a mechanism for long-term memory behavior in mammals.

When considered in isolation from available cellular/molecular neuroscientific explanations, or when considered prior to (or during) their development, psychological explanations appear genuinely causal. Let's develop the example I suggested in the previous paragraph in more detail. Kurt has just executed a series of limb movements that resulted in the successive depression of raised buttons on a telephone's vertical faceplate numbered "3" "6" "5" "4" "9" "0" "9". How might the "psychology of memory consolidation" explain this behavior?

- He rehearsed the verbal sequence fifteen times after I spoke it to him yesterday and no effective retrograde interference occurred during the thirty minutes that followed.

On its face, this sounds perfectly reasonable and perfectly causally mechanistic. Kurt's rehearsals combined with no retrograde interference caused the number sequence to be consolidated into a long-term memory accessible today. However, once we have an experimentally verified account[7] of the underlying neuronal/molecular mechanisms *tied directly to controlled behavioral data*, we can juxtapose the two "explanations" and assess their claims to revealing the event's causal-mechanistic nexus. Compare the above with:

- Following the auditory stimulus, activity in frontal neurons led to dopamine release onto hippocampal neurons that had also been activated by anatomical pathways from stimulated auditory receptors. These combined activations in thousands of selected neurons led to increased levels of freed PKA catalytic subunits in them through intracellular adenylyl cyclase-cAMP-PKA interactions, enabling some of these molecular subunits eventually to translocate to affected neurons' nuclei. There these molecules phosphorylated CREB proteins that in turn bound to CRE sites on various genes' control regions, turning on transcription, translation, and ultimately protein production. These new proteins kept the PKA catalytic subunits free from regulatory subunits and so in a persistently active state, and activated expression of late-response genes that transcribed proteins required to lastingly enhance synaptic efficacy. This process was also occurring in synapses of selected neurons downstream from the hippocampus, leading ultimately to specific motor effectors. Twenty-four hours later, inputs similar to the activation that originally induced these selective changes generated the sequence of action potentials in the motor neurons to orchestrate sequences of muscle fiber contractions and relaxations, pulling against Kurt's skeletal frame, to produce the dynamical pattern of limb movements that depressed those numbered buttons in that order on the telephone's vertical faceplate.

Of course, as the details articulated in the previous chapter reveal, even this explanation is condensed and greatly oversimplified in light of currently available knowledge.

Juxtaposed next to each other, one need not possess a detailed philosophical theory of causal explanation to judge how *explanatorily impotent*—how *empty*—the psychological causal story has become. The cellular/molecular neurobiological account explains many key causal processes that the psychological account is either completely blind to or leaves as input-output black boxes. In other words, it *explains* events that the psychological account leaves *unexplained*. When we have neurobiological causal explanations actually in place, psychological causal explanations are rendered *otioise*. No wonder there is no currently active "psychology of memory consolidation." There is no longer anything to contribute to the available causal account by working at that level. This conclusion is not a matter of "intuitions" about what constitutes a better explanation. It is rather a report of scientific practice regarding this example over the past thirty years.

This is not to say that current science eschews all mentalistic or psychological causal explanations! (On this point my new wave metascience of current cellular and molecular neuroscience *differs* from what philosophers typically understand by "eliminativism.") First, many psychological-level explanations retain their status as "causally-mechanistically explanatory" at the present time *because lower level cellular and molecular accounts are only beginning to take formative shape*. The latter are still "on promise," not yet "on offer." Impressive as it is, the "LTP is memory consolidation" accomplishment remains a parade case for current reductionistic neuroscience. And even after cellular and molecular accounts develop beyond formative stages, psychological causal explanations still play *important heuristic roles* in generating and testing neurobiological hypotheses. At the very least, they suggest where to look for the key cellular and molecular mechanisms and how best to construct behavioral paradigms that will generate useful tests of hypothesized cellular and molecular mechanisms.[8] We saw an excellent example of this second crucial heuristic role in the detailed scientific example of the previous chapter. Kandel and his colleagues constructed their crucial behavioral test—the simultaneously acquired one-trial declarative and nondeclarative learning and memory tasks—based upon well-known psychological investigations. These crucial heuristic roles of psychological concepts and explanations would seem to imply that there is something correct about them. Indeed. Their "correctness" is captured in the "rings and strings" visual metaphor in Figure 2.11 above. They are approximations of the best causal-mechanistic story we can tell at present.

There are also numerous contexts, both everyday and scientific, where the explanatory power of the cellular/molecular neuroscientific causal account isn't important and so doesn't require elaboration. Contexts do determine the answers we give and accept to "how" and "why" questions. However, it is contrary to scientific practice to say that contexts *set parameters on what count as sufficient causal explanations* (as, e.g., Horgan 2001 says). Rather, contexts determine whether a causal-mechanistic explanation, given the current state of scientific knowledge, is *genuinely causally mechanistic*. In scientific contexts, typically only one is treated as such at any given time, namely, the one genuinely available at the lowest level (and not merely "under development" or "available in principle"). This is the point I have been stressing by pointing out that there is no current "psychological study of memory consolidation," and hasn't been since its cellular and molecular mechanisms have been discovered. In everyday contexts, on the other hand, we are more lenient about the number and variety of explanations we offer and accept. This is one way that everyday practices differ from scientific ones.

At bottom, scientific practice in cellular and molecular neuroscience refuses to grant the psychological any genuine causal explanatory role *once a real neurobiological successor is firmly in place*. This lesson is apparent in the scientific example from Chapter Two. As that example illustrates, in real science the mental/psychological loses its status as genuinely causal-mechanistic once we know how the lower-level mechanisms work. Furthermore, the remaining purely heuristic roles for psychological causal explanations hardly constitute a demand for a special class of theorists—philosophers of mind—to construct a special domain of theories—theories of mental causation—to serve any useful purpose.

Finally, scientific practice at least suggests a similar fate for cellular/molecular causal mechanisms. Once biochemistry achieves causal explanations of the tertiary folding configuration of the proteins involved, and clever experimenters manipulate these proteins directly at the biochemical level to produce specific behavioral effects, molecular neurobiological explanations of the sort I presented above will also lose their (current) status as causally explanatory. They will become purely heuristic. Similarly for the future biochemistry vis-à-vis future thermodynamics or electrodynamics or whatever. If worries are beginning to churn as you contemplate the full scope of the reductionism I am urging, hold onto them. I promise to address them later in this chapter, after they emerge again when we draw consequences from our detailed scientific example for a second philosophical conundrum, the multiple realization of the psychological on the neurobiological. Before we turn to that, however, we have another "levels" question looming, this one internal to neuroscience itself.

3 WHAT ABOUT COGNITIVE NEUROSCIENCE?

3.1 "Levels" questions within neuroscience

With this talk about levels of scientific investigation and the effect that accomplished lower level explanations have on what at one time counted as causal-mechanistic (now higher level) accounts, we can expect questions about the relationship between cognitive neuroscience and cellular/molecular neuroscience. Recall from Chapter One that cognitive neuroscience is the interdisciplinary melding of psychology, neuropsychology, neuroanatomy, linguistics, computer science, and artificial intelligence. It now uses such sexy and powerful tools as "massively parallel" neural network modeling and computer simulation, functional neuroimaging, and the mathematics of dynamical systems to interpret "global" activity patterns across neuron populations. Unlike the "psychology of memory consolidation," cognitive neuroscience is currently thriving. But without question its investigations are pitched at a "higher level" than are those of cellular and molecular neuroscience.

This leads to a very interesting issue. Familiar questions about the relationship between psychology and neuroscience—in fact, the very ones we saw at work in the mental causation debate—have direct analogs concerning the levels within current neuroscience itself, and these questions are just as stark here as they are across the psychology-neuroscience divide. Consider a single example, concerning the "autonomy" of higher-level theorizing and explanation. This has been a stock question in the philosophy of psychology and cognitive science for three decades. Yet Stephen Kosslyn has recently raised this challenge within neuroscience itself, writing: "Cognitive neuroscience is a good illustration of how the whole can be more than the sum of its parts" (1997, 158). Although he admits that "cognitive neuroscience must move closer to neurobiology," he still insists that "it will not simply become neurobiology" because "cognitive neuroscience adds methods and techniques to study, and conceptualize, how the brain gives rise to cognition and behavior" (1997, 160). His "levels" relations within the brain sciences themselves will sound familiar to philosophers of psychology. Indeed, they are the very claims routinely made by autonomists about the psychological.

There are two broad approaches toward addressing levels questions. One takes its departure from the *philosophy of science*. It explicates an abstract intertheoretic relationship and applies it to questions about specific scientific levels. This approach leads to familiar theories about and disputes over relations like reduction, mechanism, supervenience, emergence, realization, and instantiation. The other approach works from *within empirical science itself*. It seeks to employ the different experimental methods and data analysis techniques used in the levels whose relation is at issue. The results are *transdisciplinary* research projects that address some phenomenon using resources from a variety of levels. The hope is that by seeing how these resources work together and interact, we will come to see how the levels relate from which they are drawn. I contend that this second approach is a much more fruitful way to address levels questions. To defend this, I'll next describe a transdisciplinary research project that my group is pursuing, indicate how we combine the methods and techniques from neuroscientific levels ranging from single cell physiology, computational modeling and computer simulation, and functional neuroimaging, show some promising preliminary experimental results, and then close by drawing some lessons that such transdisciplinary projects shed on "philosophical" disputes about autonomy versus reduction across scientific levels.

3.2 Searching for the cellular mechanisms of the sequential features of higher cognition

Neuropsychologists have known for decades that frontal cortical regions subserve many "higher" cognitive processes (Kolb and Whishaw 1996, chap. 14). Examples include language production, complex motor sequencing, planning, and problem solving. Higher cognitive processes are characteristically sequential. They proceed from one idea or representation to another in a way that seems to respect and exploit the contents of earlier steps and upcoming target steps. To grasp my half-full beer glass on the cluttered bar, behind the pretzel basket that my buddy's arm is hovering above, I must make a sequence of hand and arm movements in the proper order. To verbally utter an understandable (not to say grammatical) English sentence, I must not only get the right sounds spoken, but also in the right order. The question is, how do we do this? How do neurons in frontal regions generate the sequential aspects of higher cognition? What are the cellular mechanisms of these characteristic features?[9]

To address these questions scientifically, we followed the standard practice of first searching for a model neural circuit. Three conditions seemed obvious for a neural circuit to serve as a fruitful model in the search for the cellular mechanisms of sequential cognitive processing.

1. Its components need to be located in frontal cortex.

As mentioned above, many higher cognitive and conscious processes have been known for some time to depend on intact frontal circuits.

2. It must generate sequential outputs.

Sequential features of higher cognition are our explanatory goal. Outputs of the model circuit need not be "cognitive," but they must at least share the sequential features of higher cognition. Finally,

3. Much must be known about its cell-physiological and anatomical details.

What is lacking, and very difficult to get in alert, behaving cognizers, is knowledge of these details. A model system that meets all three of these conditions could yield a cell-physiological mechanistic account of how its sequentially structured outputs are generated that comfortably generalizes to other frontal regions that subserve higher cognition and consciousness—or at least it might suggest novel testable predictions about these.

The primate *saccade command circuit* meets all three conditions. Saccades are a type of rapid eye movement that continuously relocate the fovea (the area of the retina with highest visual acuity) on different features of a visual display. Human and nonhuman primates saccade 3-5 times per second. Most saccades are executed involuntarily, but saccades can be controlled voluntarily and consciously. The primate saccade command circuit involves both frontal and posterior parietal cortical regions. The frontal regions receive retinotopically coded visual inputs from the dorsal visual stream and project efferents to midbrain, brain stem, and neostriatum.[10] The bilateral frontal eye fields (FEF) of premotor cortex, encompassing primarily Brodmann's area 8a in humans, are prominent in this circuit. This component of the circuit thus meets condition (1) above.[11]

The role of FEFs in saccade commanding has been well studied in nonhuman primates at the single-cell electrophysiological level (Bruce and Goldberg 1985; Goldberg and Bruce 1990). Thus this circuit meets condition (3) above on a fruitful model for investigating the cellular mechanisms of the sequential features of cognitive processes. These studies show that FEFs contain neurons with pre- and post-saccadic activity. Pre-saccadic neurons have movement fields analogous to the receptive fields of sensory neurons. They fire action potentials at maximal frequency prior to saccades of a specific amplitude and direction, near-maximal frequency to saccades of closely related dimensions, and diminish to baseline firing rate prior to saccades of different dimensions. Post-saccadic neurons begin discharging action potentials over baseline rate immediately after saccade onset. Their activity also displays similar movement fields. More than 1/3 of the neurons that Goldberg and Bruce (1990) found with pre-saccadic activity also had post-saccadic activity. Their pre- and post-saccadic movement fields were directly opposed. If one of these neurons was most active prior to a saccade of a particular amplitude and direction (e.g., up and to the left), it was most active after a saccade of exactly the opposite amplitude and direction (e.g., down and to the right). Since many FEF neurons fire at different rates both before and after a single saccade, presumably the FEFs send saccade command dimensions to midbrain and brainstem circuits coded as a vector average of all active neurons (the average of each active neuron's preferred movement dimension times its spiking frequency).

Goldberg and Bruce's (1990) results using the two-step saccade paradigm suggested that the FEFs compute at least two-step saccade sequences "on the fly" or "ballistically," without the use of reflective feedback. This computation makes use of the retinotopic dimensions of the two stimuli from the initial origin encoded in pre-saccadic FEF activity and the post-saccadic activity of the cells after the first saccade. Given FEF neurons' pre- and post-saccadic movement fields, the primate FEFs appear to compute the dimensions of the second saccade by implementing *vector subtraction in eye movement space* (Figure 3.1).[12] This sequential processing in primate FEFs thus meets condition (2) on a fruitful model for the cellular mechanisms of cognitive sequential processing, at least for two-step sequences. Further evidence for the existence of multiple-step saccade sequencing comes from human eye tracking studies. When a human face is presented as a visual stimulus, normal adults respond with a characteristic pattern of saccades that move systematically to the two eyes, the nose tip, and the lips. This pattern is contrasted with that of human schizophrenics, whose saccades to the same visual stimulus are scattershot, seemingly random, and unsystematic. Schizophrenics lack the organized sequential features of normal saccadic responses. For a recent study of this effect, see Loughland, Williams, and Gordon (2002).

What about the cellular mechanisms of saccade sequences involving more than two steps? Single-cell electrophysiological work directly addressing this question does not exist, but work on two other frontal regions suggests an intriguing possibility. For more than a decade, Patricia Goldman-Rakic and her colleagues have explored activity in individual neurons in the dorsolateral prefrontal cortex (DLPFC) of rhesus monkeys during delay periods of spatial working memory tasks (Funahashi, Bruce, and Goldman-Rakic 1989; Funahashi, Chafee, and Goldman-Rakic 1993).[13] Their behavioral tasks involve saccade and anti-saccade delayed responses. Monkeys fixate on a central spot on a computer screen and a light is flashed in the periphery. They maintain fixation on the central spot for a short delay period (from 1.5 up to 6 seconds). In the saccade task, monkeys must then saccade to the remembered location of the peripheral stimulus. In the anti-saccade task, they must saccade in the opposite direction away from the remembered location of the peripheral stimulus. During the delay period, Goldman-Rakic and her colleagues record single-cell activity in DLPFC neurons that appears to hold information "on line" about the spatial location of the target. Results with the anti-saccade task verify this interpretation. DLPFC neurons that are active during the delay period of the saccade task to a specific stimulus location are also active during the delay period of the anti-saccade task to the same stimulus location, even though the monkey must make exactly the opposite oculomotor response. This indicates that activity is tied to direction

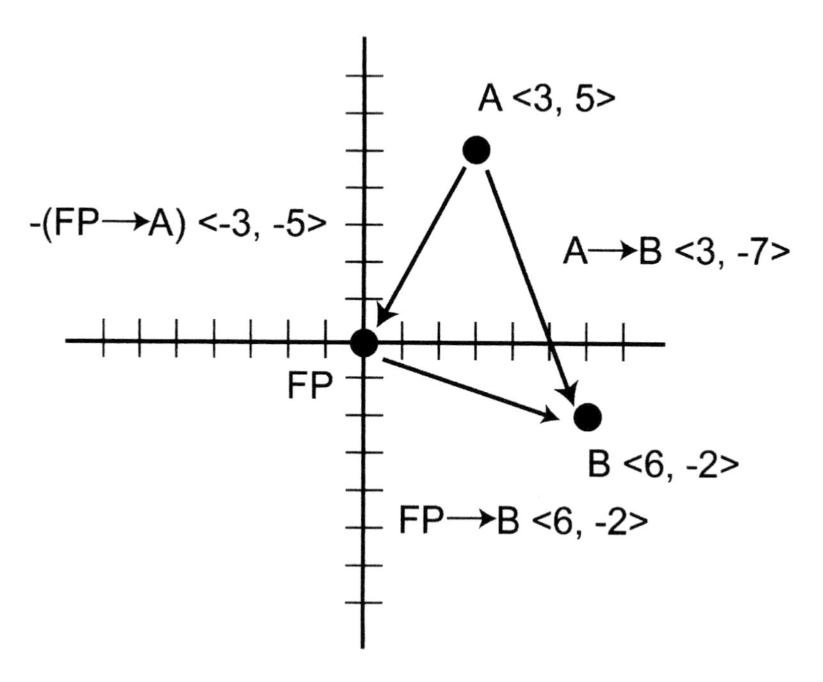

Figure 3.1. Vector subtraction in eye movement space. x-axis represents a saccade target's degrees of horizontal remove from fixation point (FP), y-axis represents degrees of vertical remove. A is the first saccade target, B is the second. In the double-step saccade paradigm employed by Goldberg and Bruce, targets A and B have both appeared and extinguished before the first saccade to A has begun (during the latency period of the first saccade in the sequence). Thus the FEFs must compute the dimensions of the second saccade from the oculomotor dimensions of the first saccade and the retinotopic location of the second target. Vector subtraction [(FP → B) – (FP → A)] computes the dimensions of A→ B: <6, -2> - <3, 5> = <3, -7>. Post-saccadic activity in FEF neurons following the first saccade provides the dimensions of –(FP → A) (i.e., (A → FP), pre-saccadic activity in FEF neurons with visual field properties provides the dimensions of (FP → B). Their summed activity yields the dimensions of A → B. See Bickle et al. (2000), especially Figures 4 and 5, for full details.

and amplitude of impending saccade; otherwise the same neuron's delay period activity would differ significantly in the saccade and anti-saccade tasks in response to the same stimulus location. Goldman-Rakic and her colleagues label this physiological activity as the "working memory fields" of individual DLPFC neurons.

More recently, Courtney *et al.* (1998) have used functional magnetic resonance imaging (fMRI) to locate a spatial working memory area specific to human frontal cortex. This area is in the superior frontal sulcus, directly adjacent to the anatomical location of the human FEFs. They measured activity increases in these regions during delay periods while humans performed a spatial working memory task, but no increase when humans performed an object identification working memory task. As is the primate DLPFC, synaptic connections between this region and FEFs are extensive.

An intriguing hypothesis combining these studies is that multiple step (> 2) saccade sequences could also be computed via vector subtraction and executed "ballistically," based on interactions between activity in FEFs and these frontal spatial working memory areas (FWMs). Pre- and post-saccadic activity in FEFs is sufficient to compute the dimensions of two-step saccade sequences. But since post-saccadic activity coding for the opposite dimensions of the saccade just executed ceases as soon as the next saccade is initiated, somehow FWM areas need to hold "on line" locations of upcoming targets in the sequence and the post-saccadic dimensions of earlier saccades. Do they? And how can this speculation about the cellular mechanisms of multiple step saccade sequences be developed and explored experimentally?

3.3 Cognitive neuroscientific resources to the rescue: Biological modeling and functional neuroimaging

Biological modeling (with computer simulation) is an ideal tool for developing testable hypotheses like the one we sought. Neurobiologist Gary Lynch and computer scientist Richard Granger nicely distinguish this type of modeling from another type that is more prominent in cognitive science and better known to philosophers:

> Recently, the question of network properties inherent in cortical design has been explored with two types of cortical simulations. The first, which might be termed abstract network modeling, employs theoretical models that use a few assumptions about neurons and their connectivities and then, through computer simulations, seeks to determine if particular, often quite complex, behaviors emerge. The biological postulates are usually simplified ... The second line of modeling research, biological modeling, seeks to exploit the rapidly growing body of information about the detailed anatomy and physiology of simple cortical networks. ... In biological models, design features are added to simulations for biological reasons only, independent of their potential computational attractiveness or complexity. (1989, 205-206).

In this spirit of biological modeling, we constructed a model of pre- and post-saccadic activity in FEFs and frontal working memory regions (Bickle *et al.*

2000, especially Figures 6 and 7). Our computational architecture and parameters were derived directly from established cellular physiology from single-cell recordings. Computations in the Vector Subtraction Core derived from FEF single-cell pre- and post-saccadic physiological discoveries by Goldberg and Bruce. Computations in the Working Memory Store derived from single-cell working memory field physiological discoveries by Goldman-Rakic and her colleagues. The model computes multiple-step saccade sequences from an initial fixation point until space in the Working Memory Store is exhausted (corresponding to time constraints of cellular activity in FWM regions) (Figure 3.2).[14]

Here, however, we confronted a common problem with biological modeling. To get realistic performance from our computer simulation (like that illustrated in Figure 3.2), we had to make some computational assumptions that could not be justified neurobiologically. A prominent assumption of this sort is the order of activity in the model's different components as the number of steps in a saccade sequence increases, e.g., from two to four. (We call this an increase in "saccade sequence burden.") Based on Goldberg and Bruce's electrophysiological data, we assumed that saccade command activity during two-step sequences is restricted to FEFs, which is realized exclusively in the computations of our Vector Subtraction Core. But as the number of saccades in a sequence increases, to three steps or more, our model requires activity in both the Vector Subtraction Core (derived from cell physiological activity in primate FEFs) and the Working Memory Store (derived from cell physiological activity in primate frontal working memory regions). There is no physiological evidence that this is the order of regional activation in the primate brain during saccade sequencing. Our model also assumed a roughly monotonic increase in the level of FEF activity as saccade sequence burden increases. Activity occurs in more FEF afferents during later steps in a saccade sequence. In addition to new visual input from the dorsal stream, FEFs also receive spatial and eye movement input from frontal working memory neurons. Again, no direct physiological evidence supports this assumption. But in the spirit of biological modeling, if we treat these assumptions as *novel hypotheses about unknown frontal neural mechanisms of saccade sequential processing*, then fMRI using Blood Oxygenation Level-Dependent (BOLD) contrast is an ideal technique for testing them empirically.[15]

One methodological advantage for functional neuroimaging of basing a study on existing cell-physiological data and a neurocomputational model derived from them is that one can simply adapt the behavioral task used by the physiologists for humans in the magnet. One doesn't need to cook up a new task from scratch. We did this with Bruce and Goldberg's (1985) double-step saccade paradigm. Six healthy adult subjects executed four cycles of

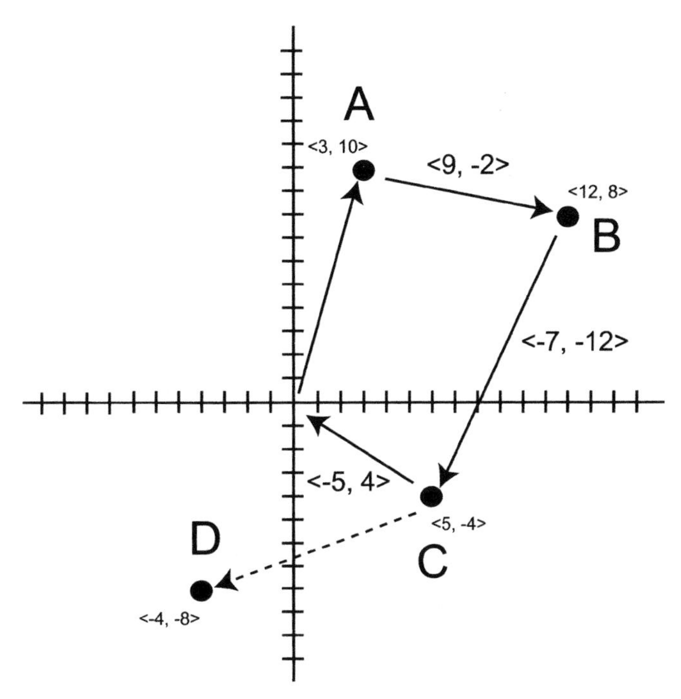

Figure 3.2. Results with a computer simulation of our neurocomputational model on a 4-step saccade sequence. Vector subtraction core first computes the dimensions of a saccade from origin to A: <0,0> occupies the previous step nodes (since this is the first saccade in the sequence) reflecting post-saccadic activity in FEF neurons, <3,10> occupies the next target nodes reflecting pre-saccadic activity in FEF neurons, so the next step nodes compute the vector sum <0+3,0+10> = <3,10> and the simulation executes a saccade of those dimension. For A→B, <-3,-10> occupies the previous step nodes, <12,8> occupies the next target nodes, so the next step nodes compute the vector sum <-3+12,-10+8> = <9,-2> and the simulation executes a saccade of those dimensions from A, landing on B (reflecting the activity of post-saccadic FEF neurons activated after the first saccade, in keeping with Goldberg and Bruce's 1990 discoveries). For B→C, <-3,-10> occupies nodes in the first layer of the working memory store reflecting working memory field activity in FWM regions, <-9,2> occupies the previous step nodes, and <5,-4> occupies the next target nodes, so the next step nodes compute the vector sum <-3+-9+5,-10+2+-4> = <-7,-12> and the simulation executes a saccade of those dimensions from B, landing on C. For C→D (dotted line), <-3,-10> occupies nodes in the second layer of working memory, <-9,2> occupies nodes in the first layer, <7,12> occupies the previous step nodes, and <-4,-8> occupies the next target nodes, so the next step nodes compute the vector sum <-3+-9+7+-4, -10+2+12+-8> = <-9,-4> and the simulation executes a saccade of those dimensions from C, landing on D. (The bold line from C back to the origin demonstrates the role of the "Return to Origin" computation and the Significance Activation Mechanism, derived from cell-physiological properties of ACC and "suppression site" FEF neurons, not discussed in this book..) Reprinted from Bernstein *et al.* 2000, Figure 8, 149, with permission from John Benjamin Publishing.

blocks of five 2-step, five 3-step, and five 4-step saccade sequencing trials, and a baseline motor task (Figure 3.3). Each target dot followed the previous one by only 100 milliseconds. We chose these timing values to insure that all stimuli were presented during the latency period of the first saccade in each sequence, to require "ballistic" processing and to place a demand on working memory capacities.[16] Each trial began when a red dot appeared for 500 milliseconds in the center of a black background. This was followed immediately by a sequence of 2, 3, or 4 yellow dots at 100 millisecond intervals, each at a randomly generated position in one of eight octants and one of two radial distances from the central fixation dot. After the final target was presented, a black screen appeared and remained until the duration of the trial. Each block consisted of five trials, each trial lasting 6 seconds, for a block duration of 30 seconds, a cycle duration of 2 minutes, and a total task duration of 8 minutes (plus an additional thirty seconds at the beginning so that subjects could acclimate to the magnet). Subjects were instructed to fixate on the red dot as soon as it appeared, and when the yellow dots appeared to saccade from one target to the next in the correct order of their appearance. They were explicitly instructed to move their eyes, not just imagine or think about where to move them. As the number of targets in a sequence increased, the task put increasing demands on subjects' working memory. Each subject reported difficulty performing the 4-step task. We used a bilateral finger-tapping motor task as a control condition. This task provides known activation in the motor strip, which we used as reference data specific to each subject.

We acquired activation data at 3 Tesla using BOLD-sensitized T2*-weighted, gradient-echo EPI. Put in its simplest terms, the BOLD signal takes advantage of the different magnetic properties of oxygenated versus deoxygenated hemoglobin. The ratio of these values gives an indirect measure of neuron (and glial cell) activity at a particular spatial resolution or "voxel size" (in this study, 3 x 3 mm), since increased cell activity requires additional oxygen carried by hemoglobin for intracellular metabolic processes (Arthurs and Boniface 2002). We acquired whole brain fMRI images in 24 slices every 1500 milliseconds for the entire task duration (8 minutes 30 seconds, so 340 fMRI data points). The first twenty images were discarded, corresponding to the first 30 seconds of fMRI data collection when a subject was acclimating to the scanner. Immediately after acquiring activation data, we performed a three-dimensional Modified Driven Equilibrium Fourier Transform whole brain scan in an axial plane to provide high-resolution anatomical images. This enabled us to co-register activation maps of individual subjects and to normalize structural images to Talairach spatial coordinates. This yields spatial resolution of structural images in three dimensions to 1 x 1.5 x 1.5 mm and provides excellent anatomical resolution and contrast between gray and white matter (Holland *et al.* 2001).

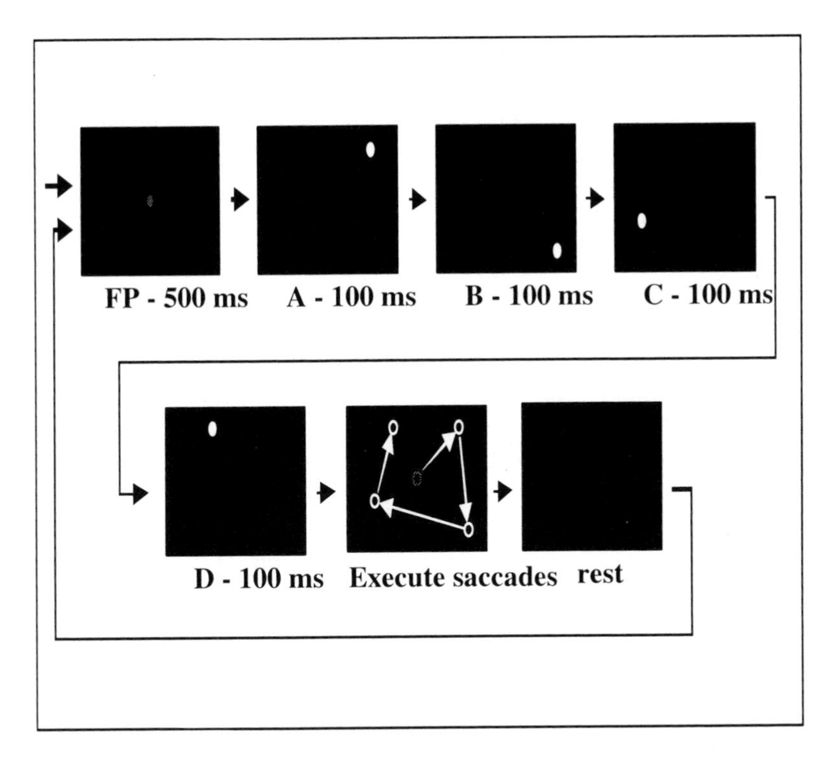

Figure 3.3. Visual display during a 4-step saccade sequencing trial. See text for discussion.

Data post-processing was done using Cincinnati Childrens Hospital Image Processing Software (CCHIPS©) (Schmithorst and Dardzinski 2000). We performed cross-correlations for each subject on a voxel-by-voxel basis between BOLD signal intensity time course and the reference function. Voxels in which the BOLD signal was above a statistically appropriate correlation threshold were overlaid on the subject's co-registered anatomical image. The resulting statistic parametric maps were then transformed into Talairach space for composite analysis, averaged across all subjects, and a composite activation map based on the average correlation value was then displayed for each comparison (e.g., 2-step versus 4-step, 3-step versus baseline task, etc.).

To identify regions subserving saccade sequential processing, we adopted two strategies. First we identified regions that were activated significantly ($p < 0.01$) during any of the saccade sequencing periods (2-step, 3-step, or 4-step) compared with the control motor task. Second, we identified voxels in the composite data where the amplitude of the BOLD-sensitized fMRI signal was correlated significantly with the number of saccade steps per trial (i.e., with increasing saccade sequence burden). These strategies clearly identified the FEFs and the two regions of frontal cortex that had previously been identified as (spatial) working memory areas, namely, a region in DLPFC and another around the superior frontal sulcus.[17] Voxel-by-voxel correlations between the BOLD signal and number of saccade steps yielded composite activation maps. We transformed correlation values into a t-statistic and averaged across all subjects. Voxels that exceeded a significance of $p < 0.01$ were overlaid across the averaged anatomical data set (Bickle *et al.* 2001, Figure 5).

We also plotted time courses for each subject for five Regions of Interest (ROIs): FEFs, DLPFC, superior frontal sulcus, posterior parietal cortex, and anterior cingulate cortex. For each ROI, we graphed normalized BOLD signal values reflecting level of activation against fMRI data point in the block design. Because the BOLD signal changes over time due to brain blood flow that has nothing to do with the task, we "detrended" (also known as "drift corrected") each time course to filter out this low frequency component and remove any baseline drift. The procedure is to fit a quadratic function to the data from each voxel and then subtract the linear and quadratic components from the measured time course data.[18] To account for factors like hemodynamic and attention lag, we also threw out the first five data points (frames) for each epoch. (All this is standard fMRI image analysis.) Using this corrected data, we then computed composite activity for each ROI during 2-step, 3-step, 4-step, and baseline task performances (Figure 3.4).

These results suggest numerous conclusions about the biological plausibility of our neurocomputational model of saccade sequential processsing. First, our model is based on cell-physiological properties of regions that are indeed active during saccade sequencing: the FEFs and two frontal working memory areas.[19] Second, FEF activity does increase in monotonic fashion with increasing saccade sequencing burden, from the 2-step task to the 3-step and from the 3-step to the 4-step (Figure 3.4A). Finally, activity in frontal working memory regions also increases with saccade sequencing burden, and our data suggests a dissociation between DLPFC and superior frontal sulcus activation. Activity increase in DLPFC is most prominent in the shift from 2-step to 3-step saccade sequences, while activity increase in superior frontal sulcus is most prominent in the shift from 3-step to 4-step sequences.[20] As noted above, given the speed of stimuli presentations, the 4-step task places a heavy burden on working memory mechanisms. All these results suggest that some purely computational assumptions of our neurocomputational model do have biological plausibility when treated as testable novel hypotheses about the cellular mechanisms of saccade sequential processing in frontal circuits.

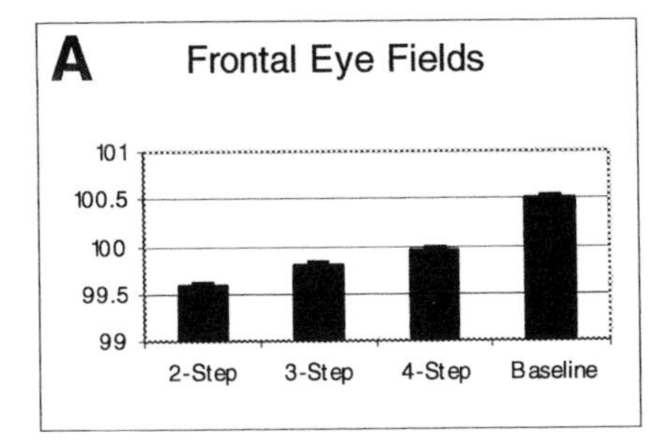

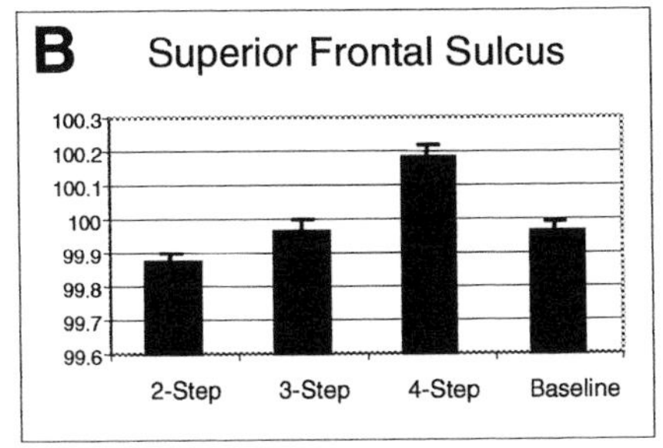

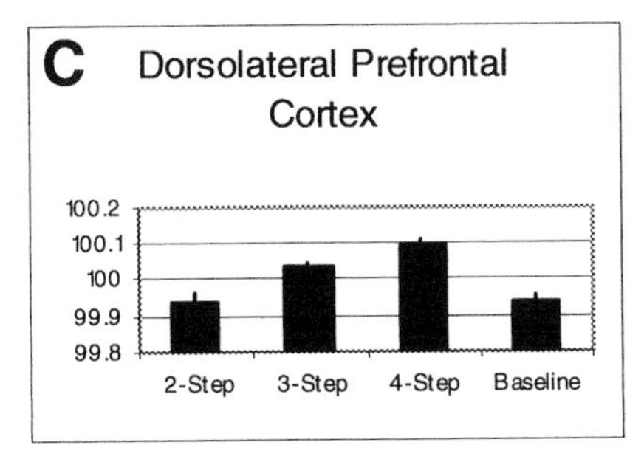

Figure 3.4. "Detrended" ("drift corrected") normalized composite mean BOLD signal intensity in three regions of interest during 2-step, 3-step, and 4-step saccade sequencing and control finger tapping trials. See text for discussion.

Data from this preliminary study also reveal an experimental design flaw; but upon analysis, results from this flaw serendipitously help to verify the biological plausibility of our model of saccade sequencing. Consider the high level of FEF activity during the baseline motor task (Figure 3.4A above). What accounts for this? During that finger-tapping task, subjects were staring at a blank black screen. Guess what humans (and other primates) do when they stare out into darkness? They saccade at roughly the normal rate, 3 times per second. This explains the high level of FEF activity in control as compared to experimental tasks. In the latter, subjects are forced to make only a limited number of saccades, after which they return their eyes to the center of the screen, anticipating the next saccade sequencing trial.[21] But consider also the lower level of activity in the frontal working memory regions during the baseline task (Figure 3.4B, C). This suggests that although subjects were saccading frequently during it, their saccades were not guided by working memory information about either stimulus spatial location or oculomotor dimensions of recently executed saccades. In other words, saccades during the control task were random, one-step saccades, not organized sequences. So saccade *sequencing* (beyond two steps) does seem to require interactions between FEFs and frontal working memory region activity, just as our model hypothesizes.

3.4 Philosophical lessons from transdisciplinary neuroscience

What lessons do this transdisciplinary cell-physiological, biological modeling and neuroimaging study suggest for the "big" issues I raised at the beginning of this section about the relationship between cognitive and cellular neuroscience? Bear in mind that figuring out the cellular mechanisms of *saccade* sequential processing is not our ultimate explanatory goal. We chose to model and explore the frontal saccade command circuit because it itself was a *model* circuit for exploring the cellular mechanisms of sequential cognitive processing. Expansion of the frontal lobes is widely believed to underlie many of the higher cognitive abilities that distinguish humans from other primates. However, the neuronal circuitry across much of the human frontal lobe remains poorly understood at the cell-physiological level, due in part to technical limits on studying human brain function and to the difficulty of identifying and characterizing the higher cognitive functions. The usual procedure in science, when confronting these kinds of difficulties, is to focus on a simpler system that shares characteristic features of the processes in question but whose components are more readily accessible experimentally. The primate saccade command circuitry provides exactly this sort of model system for the sequential properties of higher cognition (including our streams of conscious experiences). Saccading is neither typically "cognitive" nor conscious (although it can be under cognitive and conscious control), but it is sequential. By coming to understand the cell-physiological mechanisms by which saccade sequences are computed and executed, we seek to uncover *testable hypotheses* about the cellular mechanisms of the sequential aspects of higher cognition. One hypothesis resulting from our pilot study is that sequential processing (cognitive or not) is vector subtraction implemented neurally. It is now commonplace in cognitive neuroscience to interpret activity across a population of neurons in terms of vector space representations and transformations (Churchland and Sejnowski 1992). Vector subtraction computes the dimensions of future paths through the appropriate space, and is realized neurally by the interactions between pre-movement, post-movement, and working memory activity in the individual neurons comprising the population (Bickle *et al*. 2000).

This judgment is strengthened by two features specific to our transdisciplinary study. First, cytoarchitecturally—with regard to cell types and distribution, and both laminar and columnar structure—the frontal saccade command regions are composed of "standard frontal cortex with a distinct granule layer." They thus possess exactly the cellular resources as most of frontal cortex (Parent 1996). These similarities suggest that the computational strategies implemented in the cell properties and connectivities of the primate saccade sequencing system—vector subtraction with an interactive working memory store—are available to many other frontal cortical regions known to subserve sequential cognitive processing. Any cognitive process that can be characterized mathematically as a pathway through a multi-dimensional vector space could be implemented neurally via vector subtraction by the appropriate set of pre-vector, post-vector, and working memory field neurons. The scope of this way of "mathematizing" brain function in contemporary cognitive neuroscience (to say nothing of its applications in "connectionist" cognitive science more generally) suggests testable empirical hypotheses for physiological investigation. If the FEFs and frontal working memory regions implement saccade sequential processing in this fashion, the cytoarchitectural similarities with the rest of frontal cortex suggest that the latter might implement it in similar cellular mechanisms. Second, the continued biological justification that the fMRI studies give to what were purely computational assumptions of our model increases the plausibility of using the model to generate testable hypotheses about the cell-physiological mechanisms of the sequential features of other frontal cognitive and conscious processes. For a given sequential cognitive process, representable mathematically as trajectories through a vector space, can single-cell investigations find neurons with the appropriate pre-activity, post-activity, and working memory fields to implement iterated vector subtraction (in the fashion that the saccade command circuit appears to)? What single-cell neurophysiologist wouldn't like such leads about neuron response properties to look for during cognitive processing tasks?

To summarize the discussion so far: saccade commanding and execution are (typically) neither cognitive nor conscious, but their outputs are sequentially organized. In addition, these components and circuits are well understood at the single-cell physiological level, located in frontal cortex, and have been characterized by a successful neurocomputational model, some of whose purely computational assumptions have now been verified biologically. (Our transdisciplinary research project demonstrates the last two points.) Like any fruitful scientific model, ours suggests testable hypotheses for future research toward uncovering the cellular mechanisms of the sequential aspects of higher cognition and conscious experience. Can we be sure that our model has focused upon the essentials of even this single feature of cognition? Can we be sure that we aren't being misled by the cellular mechanisms of a simpler system producing sequential outputs? Can we be sure that other strategies for uncovering the cellular mechanisms of higher cognition won't be more successful (e.g., Bechtel and Richardson's 1993 "decomposition and localization" strategy or Craver and Darden's 2001 strategy for "discovering mechanisms")? Our answers are no, no, and no. But these negative answers are for the unexciting reason that we are doing science, where no algorithm for successful discovery exists.[22]

What about cognitive to cellular neuroscience reduction versus the "autonomy" of higher level methodologies, explanations, and theories? Transdisciplinary projects employing methods and results from a variety of levels (like ours) address this issue empirically. We use neurocomputational modeling and functional neuroimaging to answer questions that it is difficult to imagine addressing solely at the single-cell physiology level. If that is all that "autonomy of the higher level" amounts to, then we treat neurocomputational modeling functional neuroimaging as "methodologically autonomous" from cell physiology. But that typically isn't all that proponents of "higher level autonomy" want, at least among philosophers. They want autonomy of *theory* or *explanation*; they claim that mechanisms uncovered by higher level investigations are "independent" of the lower-level details. That sort of autonomy is clearly no part of transdisciplinary science-in-practice, at least of the sort we pursue. We use the higher level methodologies, resources, and techniques as a way to get to the cellular mechanisms of cognitive processing. And I contend that our approach is characteristic of most real transdisciplinary neuroscience, as opposed to philosophers' and cognitive scientists' fantasies about how it proceeds. For us, higher level theories and methods have a useful and seemingly ineliminable *heuristic* role to play in the search for lower level mechanisms and reductions. They greatly increase the descriptive base for lower level experiment and theorizing. They tell us where in the brain to look and they guide us in constructing behavioral tasks that isolate the crucial dependent and independent variables. *But that is all they do, and all they can do.* When they've exhausted this descriptive and methodological function they fall away, much like Wittgenstein's ladder.[23] Higher level accounts reduce to lower level mechanisms, after the former function to help us discover the latter. If you think that they do more, then your intuition is no part of transdisciplinary research in mainstream neuroscience and you really are a "levels" dualist, whether or not you admit to it. Features "of the whole," "of the system," "of the population," are discovered to be nothing but complex sequences, combinations, and interactions that occur between the individual components, specifiable ultimately in lower level terms and interactions. What else could they be?

Nothing prevents us from resolving more general levels questions across cognitive science and neuroscience in the same way. Cognitive scientific investigations, methodologies, theories, and explanations are essential heuristics in the search for lower level neuronal mechanisms. They generate data, descriptions, and behavioral tasks as part of transdisciplinary research projects that it is difficult to imagine gaining from purely cellular and molecular investigations. This much "methodological autonomy" is an ineliminable part of standard scientific practice. But that is all the "autonomy" that transdisciplinary scientific practices warrant, and that much is consistent with ruthless reductionism.

4 PUTNAM'S CHALLENGE AND THE MULTIPLE REALIZATION ORTHODOXY

"It is remarkable that the cAMP signal transduction pathway, including its nuclear components, seems to be required for memory-related functions in each of these species and behavioral tasks" (Yin *et al.* 1994, p. 55).

I mentioned in Chapter One (section 4.1) that one of the most influential challenges to psychoneural reduction in the philosophical literature rests upon *multiple realization*. There we saw ways that reductionists have tried to undermine anti-reductionist conclusions urged from it. Mainly, reductionists have granted the multiple realization premise and argued that anti-reductionist conclusions drawn from it are invalid. Few have challenged the truth of multiple realization. It is time to see why, and in light of the detailed example from the previous chapter and additional scientific details hinted at in the quote that prefaces this section, to construct this more direct assault. I am convinced that this is the definitive reply to one of the most influential arguments in late-20th century philosophy.

Multiple realization in philosophy of mind begins explicitly with Hilary Putnam. In defense of the "functionalist" view of mind that he developed and championed throughout the 1960s, Putnam laid down a challenge to his chief competitors: the early "central state materialists" who advocated identifying mental with neural types (properties, states, events). Using "pain" as his paradigmatic mental type, Putnam writes:

> Consider what the brain-state theorist has to do to make good his claims. He has to specify a physical-chemical state such that any organism (not just a mammal) is in pain if and only if (a) it possesses a brain of a suitable physical-chemical structure; and (b) its brain is in that physical-chemical state. This means that the physical-chemical state in question must be a possible state of a mammal's brain, a reptilian brain, a mollusk's brain (octopuses are mollusca, and certainly feel pain), etc. (1967, p. 45)

Since the brain-state theorist makes the same claim about every mental state, his liability is even greater. Putnam continues: "If we can find even one psychological predicate which can clearly be applied to both a mammal and an octopus (say, "hungry"), but whose physical-chemical correlate is different in the two cases, the brain state theory has collapsed" (ibid.). It is "overwhelmingly likely," Putnam asserts, that we can find such a state—many, actually.

Thus *multiple realization* entered into the philosophy of mind. The premise asserts that a given psychological kind (property, state, event) is realized by distinct physical kinds. Providing a precise definition of 'realization' has proved difficult. Must it be a necessary truth that the realizing state, property, or event obtain only if the realized state, property or event obtain, or is contingency enough? If necessity is required, which strength of necessity is sufficient (physical, metaphysical, logical)? Despite these controversies, multiple realization has achieved consensus as a true and crucial premise in an argument against mind-brain identity theory and psycho-physical reduction. This status has remained even as functionalism, the view that spawned the argument, has given way to nonreductive physicalism. One important reason for its staying power, given the concerns in this book, is that neuroscience itself seems to provide examples of creatures whose behavior is describable using the same psychological concepts but whose nervous systems are very different. The realizing neural states across the creatures would not be identical. Hence multiple realization seems verified empirically.

In the remainder of this chapter, I aim to challenge the truth of multiple realization as it applies to creatures here on earth. From the perspective of behavioral and systems neuroscience, multiple physical realizations of shared psychological kinds seem obviously to obtain. But as neuroscience's core over the past two decades has shifted, increasingly to the level of molecular manipulations and investigations, multiple realization at the systems level gives way to evolutionarily conserved, shared mechanisms across otherwise vastly different species. The molecular mechanisms determining neuron activity and plasticity are the same in invertebrates through mammals. In light of these discoveries and the direct ties that have been forged to behavioral effects, the emerging "links" between molecules and mind cast doubt upon multiple realization. I will flesh out this argument by supplementing the scientific details from the previous chapter with additional ones from studies on invertebrates, in particular on fruit flies and sea slugs. These recent discoveries answer Putnam's challenge. As "overwhelmingly unlikely" as it might seem from the philosopher's armchair (and the systems neuroscientist's laboratory bench), cellular and molecular neuroscientists are discovering "physical-chemical states" that serve as shared mechanisms for shared psychological events across biological phyla. And principles of molecular evolution suggest that more of these will be discovered as these sciences progress. Multiple realization, meet molecular neuroscience.

Some observations and caveats are immediately in order. First, Putnam's original challenge was an *empirical scientific* challenge. He claimed that (circa the empirical data and theories of 1967) unitary physical-chemical states were unlikely to be found in the variety of earthly organisms for which it seems reasonable to assume identical psychological events. To address his challenge, one must show that science is finding such unitary physical-chemical states. One can only show this by presenting scientific details. Philosophers who reject the relevance of empirical science will not be impressed. But they've probably long since put down this book, and they should equally not be impressed by Putnam's challenge and its anti-reductionist conclusion. Putnam's challenge rests directly on intuitions about scientific facts and what empirical investigations will or will not discover.[24]

Unfortunately, over the three and one-half decades since Putnam first issued his challenge, its empirical grounding has fallen out of focus. Anti-reductioist philosophers nowadays typically speak of multiple realiz*ability* and lace their discussion with "thought experiments" involving fantasized cognizers like silicon-based aliens and artificially intelligent electronic robots. This is because most philosophers assume that identity holds across "all possible worlds," either necessarily or as a matter of scientific law. Putnam (1967) himself strapped the "brain-state theorist" with this additional burden, writing that any proposed physical-chemical state must also be "a state of the brain of any extraterrestrial life that might be found that will be capable of feeling pain" (1967, p. 45). Jerry Fodor (1974) exploited this "nomological" assumption in his important elaboration and extension of Putnam's original argument. This explains the popular "thought experiments" in the multiple realizability literature of beings that share our psychological kinds but lack our organic brains at any level of physical description.

This broader sense of multiple realiz*ability* and philosophers' "possible world" fancies do not concern me. I don't know whether identity holds across "all possible worlds," or even across all "physically possible worlds." I don't know the "conceptual" or "nomological limits" of our psychological concepts. But I take comfort in the fact that you don't, either, regardless of the strength of your intuitions. Jerry Fodor once remarked that

> self-confident essentialism is philosophically fashionable this week. There are people around who have Very Strong Views ('modal intuitions,' these views are called) about whether there could be cats in a world in which all the domestic felines are Martian robots, and whether there could be Homer in a world in which nobody wrote the Odyssey or the Iliad.

> Ducky for them; their epistemic condition is enviable, but I don't myself aspire to it. (1987, 16)

I concur completely and apply this attitude to the "possible world" scenarios that transform multiple realiz*ation* from an interesting empirical problem into multiple realiz*ability*, a "conceptual puzzle" about "the scope of our psychological concepts." As you might have already gathered from my neo-Carnapian outlook sketched in Chapter One, I steer clear of pragmatically fruitless questions. I'll worry about brainless yet pained or belief-entertaining aliens and robots as soon as one crosses my path. My concern is with existing earthly creatures. If the scope of my concern is too narrow for your philosophical sentiments, so be it. Scientists don't give a hoot for philosophers' Very Strong Modal Intuitions about kind identity across possible worlds, and their enterprises are doing just fine.

I'll make one more remark on this point and then move on. Since the heyday of "central state materialism" in the 1950s and early 1960s, "brain-state" theorists have emphasized the *contingent*—read: this worldly—nature of their identity or reductionist claims. No serious "brain-state" theorist over the past forty years has claimed that neural expressions will provide synonyms for psychological expressions, any more than the "lightning is atmospheric electron discharge" hypothesis claims to. It might be fashionable in post-Kripke philosophy to insist that all identity claims hold necessarily, but fortunately scientists don't bother reading Kripke and keep right on making and testing identity claims that purport to hold in the real world. Despite this unfortunate detour into pragmatically fruitless metaphysics that the multiple realization issue took, however, there still remains within it a genuine and unanswered empirical challenge. Answering this challenge is my goal.

The next point to note at the outset is that no paradox is looming in my project. That molecular neuroscientists are finding shared mechanisms that underlie *shared* psychological kinds across species *does not* imply that all species possess similar psychological capabilities. That implication would be a *reductio ad absurdum* for the ruthless reductionism of current mainstream neuroscience. Any account that leaves out the psychological differences between, say, humans and sea slugs, will be seriously incomplete as an account of cognition and behavior! Fortunately, nothing is implied or suggested in the account about to be presented about vast psychological *differences* across species. The multiple realization challenge only speaks to *shared* psychological kinds. Obviously, psychological differences must have distinct mechanisms of some sort.

Current cellular and molecular neuroscience does have a going story about these differences. This story has been characterized by Richard

Hawkins and Eric Kandel as a "cell-biological alphabet" out of which different "words" and "sentences" are constructed to explain different psychological capacities.[25] They write:

> Do the [intracellular] mechanisms so far encountered form the beginning of an elementary cellular alphabet? That is, can these units be combined to yield progressively more complex learning processes? We would like to suggest on theoretical grounds that such an alphabet exists and that certain higher-order forms of learning generally associated with cognition can be explained in cellular-connectionistic terms by combinations of a few relatively simple types of neuronal processes. (1984a, 386)

They draw their cell-biological letters out of earlier research on the sea slug, *Aplysia californica*, but argue for their applicability to cognitive learning in mammals:

> We propose that higher forms of learning may utilize the mechanisms of lower forms of learning as a general rule; and second, we speculate that this may occur because higher forms of learning have evolved from lower forms of learning. ... Thus, whereas individual neurons may possess only a few fundamental types of plasticity that are utilized in all forms of learning, combing the neurons in large numbers with specific synaptic connections (as occurs, e.g., in mammalian cortex) may produce the much more subtle and varied processes required for more advanced types of learning. (1984b, 391)

However, I mention this approach merely to show that no paradox is looming in my reply to multiple realization. Given that its challenge focuses on shared psychological kinds, a discussion of multiple realization is not the place to air complaints about this approach toward explaining psychological differences.

Finally, it is impossible (obviously) to establish the unique realization of each type of psychological state across all the species that possess it. At present, we don't know enough about the underlying neural mechanisms for many types, especially the molecular mechanisms where I am claiming that the unique realizations lie. Still, if we can find one prominent shared psychological kind that *appears* to be realized differently in the nervous systems of different species possessing it, but which turns out actually to be uniquely realized by shared molecular mechanisms, this single empirical example would bolster a general hypothesis. This argument will be even stronger if that example illustrates a principle, the evolutionary conservation of molecular mechanisms, that holds generally, on independent grounds. I contend that such an example already exists in current molecular accounts of memory consolidation.

We are about to embark again into detailed molecular neuroscience. The sledding will be no easier here than it was in Chapter Two, although readers quickly will be struck with a feeling of déjà vu. ("Haven't we heard of these molecules, signaling pathways, and effects on long-term memory behavior before?") But then, did you really expect a challenge with the staying power of multiple realization to yield to a *simple* empirical reply?

5 MOLECULAR MECHANISMS OF NONDECLARATIVE MEMORY CONSOLIDATION IN INVERTEBRATES

5.1 Single-gene fly mutants for associative learning

In the mid-1960s, Seymour Benzer introduced a technique for generating chemical mutations of single genes. This work drew upon the beginnings of biotechnology that has since mushroomed into the techniques used today to manipulate the mammalian genome (discussed in Chapter Two, sections 4.2 and 5.2 above) The fruit fly, *Drosophila melanogaster*, proved to be an excellent experimental preparation. Benzer's group developed an olfactory shock-avoidance conditioning procedure for *Drosophila* (Quinn *et al.* 1974) and soon produced the first single-gene learning and memory mutant, *dunce* (Dudai and Quinn 1976). Quinn continued this mutagenesis approach with shock-avoidance procedures and produced four other *Droso-phila* learning and memory mutants. Other training and testing procedures for *Drosophila* were developed, including several operant procedures (Connolly and Tully 1997). At present, more than twenty fly learning and memory mutants have been identified (Dubnau and Tully 1998).

A common conditioning procedure begins by trapping a group of flies in a chamber. Training involves exposing them to odor A (the CS^+) paired with a shock (US) through the chamber's landing apparatus. This is followed by exposure to odor B (the CS^-) that is not paired with a shock. "Spaced" training intersperses training sessions with rest periods. "Massed" training presents sessions consecutively without rest intervals. Trained flies are then transferred to the choice point in a T-maze where they "choose" between odor A (CS^+) in one arm and odor B (CS^-) in the other, presented simultaneously. Greater than 90% of wild-type (non-mutated) flies avoid the CS^+, while less than 5% avoid the CS^- (Tully and Quinn 1985). Spaced training with 10 training sessions produces long-term memory for the CS$^+$-US pairing in flies that is measurable for more than seven days. This conditioning assay has been used to "dissect" the specific learning and memory deficits of induced *Drosophila* mutants, e.g., by varying the nature of the training sessions and the time between training and T maze choice test.

Biochemical analysis of *dunce* and *rutabaga* mutants suggests that the intracellular cyclic adenosine monophosphate (cAMP) "second messenger" pathway is crucial for *Drosophila* olfactory learning and memory (Levin *et al.* 1992). In a series of recent experiments, Tully's group has elaborated the role this pathway plays in *Drosophila* memory consolidation. They cloned a *Drosophila* gene, *dCREB2*, which transcribes a number of related protein products (Yin *et al.* 1995b). One, dCREB2-a, is a protein kinase A (PKA)-responsive transcriptional activator. Another, dCREB2-b, is a repressor of PKA-responsive transcriptional activation. Readers of the previous chapter are familiar with the role of the related mammalian genes and their protein products in postsynaptic E-LTP, L-LTP, and declarative long-term memory consolidation.

Using germline transformational techniques standard in invertebrate molecular biology for two decades, Tully's group generated *Drosophila* mutants that overexpress *dCREB2-b* under the control of a heat shock promoter (hs-dCREB2-b) (Yin *et al.* 1994). This promoter is only turned on to transcribe the transgene's protein product after mutant adult flies are exposed to heat shock. Experimental mutants get exposed to heat shock; control mutants do not. This is a standard practice to control for possible developmental effects of the transgene's insertion that might affect behavior but are not specific to learning and memory. When expressed, the mutant transgene guides the production of an overabundance of dCREB2-b, the CRE repressor. This blocks the expression of various genes induced by the cAMP-PKA cascade. Using the Tully and Quinn (1985) olfactory association procedure (described above), Yin *et* al. (1994) showed that inducing the *hs-dCREB2-b* transgene disrupted long-term memory in mutant flies after spaced training. Induced *hs-dCREB2-b* mutants displayed significantly poorer retention of the CS$^+$-US association when tested at both 24 hours and seven days later, compared to both wild-type flies and *hs-dCREB2-b* mutant controls (flies with the transgene inserted that were not exposed to heat shock, and so in which the transgene was not expressed). Despite this long-term memory deficit, however, induced *hs-dCREB2-b* mutants displayed normal initial learning, olfactory acuity and shock reactivity (all assayed using standard techniques). On a massed training regime that produces 3-hour but not 24-hour retention in wild-type flies, induced *hs-dCREB2-b* mutants performed similarly to both wild-type and non-induced *hs-dCREB2-b* mutant controls. Clearly, overexpressing the *hs-dCREB2-b* transgene produces an effect specific to memory rather than perception and selectively disrupts consolidation of long-term olfactory conditioning.

In a subsequent study, Tully's group inserted and expressed another heat-shock promoted transgene, *hs-dCREB2-a*. This mutation overexpresses dCREB2-a, the activator isoform of the protein. In wild-type and non-induced (non-heat shocked) *hs-dCREB2-a* mutant controls, optimal long-term memory (tested seven days after training) in the Tully and Quinn T maze requires 10 spaced training sessions. Induced *hs-dCREB2-a* mutants, however, showed comparable long-term retention after only a single training session. Again, using standard assays, olfactory acuity and shock reactivity were normal in the induced *hs-dCREB2-a* mutants compared to wild-type and non-induced *hs-dCREB2-a* mutant controls. This result indicates that this transgene's overexpression also produces specifically a memory effect. Finally, long-term memory formation was not enhanced (compared to controls) in transgenic flies expressing a mutant CREB isoform that is resistant to phosphorylation by PKA. Based on their combined results from these studies, Yin *et al.* conclude that "opposing functions of CREB activators and repressors act as a "molecular switch" ... required to form [long-term memories]" in *Drosophila* olfactory conditioning (1995a, p. 110).

Figure 3.5 illustrates the current model of memory consolidation in *Drosophila*. Neurons in the mushroom body receive sensory inputs from both the olfactory and the foot shock sensory pathways. The latter come in through a modulatory neuron releasing a catecholamine as its neurotransmitter (dopamine, DA, or serotonin, 5-HT). The CS^+ pathway activated by odor A generates action potentials in specific mushroom body neurons and a subsequent rise in calcium ion influx (Ca^{2+}). (This is a standard neuronal effect I described in Chapter 2, section 3.2.) Neurotransmitter release by the presynaptic olfactory neuron generates action potentials in the postsynaptic mushroom body neuron. As the action potential traverses the length of the mushroom body neuron's axon (and backpropagates to the soma and dendrites), it opens voltage-gated Ca^{2+} channels and Ca^{2+} flows into the cell through these opened gates by the forces of diffusion and electrostatic pressure. Intracellular Ca^{2+} interacts with calmodulin (CaM), producing increased amounts of Ca^{2+}-CaM complex. Adenylyl cyclase molecules have a Ca^{2+}-CaM complex binding site; binding at these sites primes these molecules for subsequent intracellular interactions. In the paired CS^+-US condition, catecholinergic release by the modulatory interneuron in the US pathway occurs during this rise in intracellular Ca^{2+}, Ca^{2+}-CaM complex, and adenylyl cyclase priming. The DA or 5-HT receptor on the mushroom body neuron is metabotropic, coupled to a G protein complex that also activates adenylyl cyclase. The combined results of CS^+-US pairing makes adenylyl cyclase molecules in specific mushroom body neurons more efficient at converting adenosine triphosphate (ATP) into cAMP. Thus in these specific neurons, more cAMP molecules are available to bind to sites on the regulatory subunits of more cAMP-dependent protein kinase A (PKA) molecules. cAMP binding frees up more PKA catalytic subunits. In the short term, more freed PKA catalytic subunits phosphorylate more potassium (K^+) channels in the neuron membrane, closing these channels and preventing K^+ efflux. The result is broadened action potentials, with subsequent increases in the amount of neurotransmitter released by specific affected neurons. PKA catalytic subunits are also thought to facilitate neurotransmitter release by direct actions on release machinery in presynaptic active zones (by a mechanism still being investigated). This increase in neurotransmitter release strengthens responses in other specific neurons further downstream in the motor output pathway, leading to odor avoidance (behavior). Mushroom body neurons in the CS^- pathway, which aren't followed by the US and subsequent DA or 5-HT binding, lack the combined step that renders adenylyl cyclase molecules more efficient at ATP-cAMP conversion. Events further downstream involving these neurons, and ultimately behavioral effects (odor avoidance), are reduced substantially. All this molecular neuroscience should be familiar, since these are the molecular mechanisms of mammalian E-LTP, only here transposed into a presynaptic key. (Compare Figures 2.4, 2.6 and 3.5.)

Familiarity deepens when we consider the effects of repeated CS^+-US pairings over the appropriate time period (as occurs, e.g., in Tully and Quinn's 1985 multiple session "spaced" training). The repeated paired stimuli continually increase the concentration of freed PKA catalytic subunits in presynaptic terminals of specific mushroom body neurons. Eventually this reaches levels high enough for catalytic PKA subunits to translocate to the neuron's nucleus. There these molecules phosphorylate the CREB isoforms. During the rest intervals of spaced training, the net function of phosphorylated CREB activators comes to exceed that of CREB repressors, initiating a cascade of immediate early gene expression. Phosphorylated dCREB2-a molecules bind to cAMP response elements (CRE sites) on these genes' control regions, initiating transcription. This initiates translation of new proteins that yield long-lasting changes to structure and function of presynaptic terminals of specific mushroom body neurons, namely, increased amounts of neurotransmitter released by individual action potentials and more active zones (release sites). The result is increased efficacy at specific mushroom body-follower neuron synapses in the motor output/odor avoidance pathway. Ultimately this increased efficacy produces changes in motor output. Since this plasticity in neurotransmitter release rates is driven by new gene expression and protein synthesis, it is much more long-lasting than the transient, short-term changes that are localized to synapses and independent of new gene expression. Quite literally, the procedures that "consolidate" long-

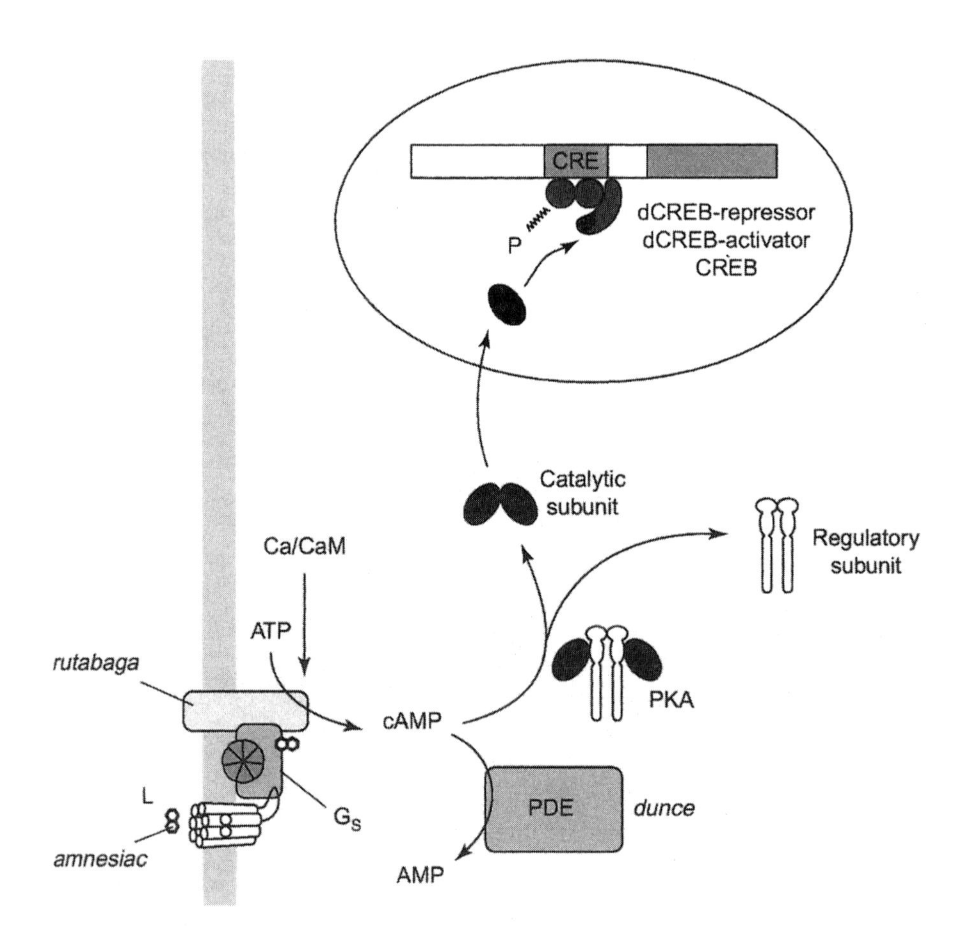

Figure 3.5. Molecular mechanisms of synaptic facilitation in *Drosophila*. Presynaptic facilitation via these molecular pathways leads to enhanced excitatory neurotransmitter release onto postsynaptic follower neuron, and hence increased postsynaptic activity to afferent stim-uli. Reprinted from *Trends in Genetics,* **15**, "Genetic approaches to memory storage," 463-470, Copyright 1999, with permission from Elsevier Science.

term memory in *Drosophila* generate permanently altered neurons in terms of their fundamental protein make-up. Previously dormant genes get expressed and new protein products result. The key step is the availability of PKA catalytic subunits in numbers sufficient to translocate to the cell nucleus. As we are beginning to see, this is a common theme across biological classes and types of memory.

5.2 Consolidating nondeclarative memory in the sea slug, *Aplysia*

Work with induced *Drosophila* single gene mutants was a crucial first step toward discovering both the molecular-genetic and biochemical aspects of learning and memory. But measuring changes in fruit fly neurons and circuit organization presents daunting technical challenges. With its relatively simple nervous system, readily accessible neurons, and more easily measured behavioral repertoire, the marine invertebrate *Aplysia californica* (a sea slug) has also been a central experimental preparation in cellular and molecular studies of learning and memory for one-quarter century. Indeed, some of the mechanisms of the current *Drosophila* model were first discovered experimentally in *Aplysia*. *Aplysia* research has also yielded increased knowledge about many molecular details first discovered in *Drosophila*.

For example, it is now clear from *Aplysia* research that the cellular and molecular mechanisms of classical conditioning, both short-term and long-term, are elaborations on those of sensitization, a simpler form of learning and memory. Behaviorally, sensitization is a heightened responsiveness in an organism's entire range of defensive reactions following a noxious stimulus. In *Aplysia*, the gill withdrawal reflex following a tail shock has been a common laboratory preparation (Frost *et al.* 1985). In normal contexts, a weak tactile stimulus to *Aplysia* siphon (e.g., touching the fleshy spout lightly with the tip of a paintbrush) produces a moderate gill withdrawal that quickly habituates with repeated stimuli (moderate in terms of the speed of the gill withdrawal into the mantle cavity and the amount of gill withdrawn). Following a single electric shock to the tail, however, *Aplysia* withdraw their gill more quickly and completely into the cavity to the weak tactile stimulus. This effect persists for minutes after a single shock and can be extended with additional shocks (making this preparation another laboratory analog of stimulus repetition). Five or more shocks presented in spaced training yields sensitization of gill withdrawal that persists for two or more days, up to two weeks (Bailey and Chen 1989).

In terms of its cellular organization, this *Aplysia* sensitization circuit begins at sensory neurons innervating the siphon and others innervating the tail. Siphon sensory neurons synapse directly on gill motor neurons and on interneurons that also synapse on gill motor neurons. Tail sensory neurons synapse on a special class of modulatory interneurons that in turn synapse on siphon sensory and gill motor neuron soma (cell bodies) and presynaptic terminals. Sensitization starts with activity in these modulatory interneurons driven by the tail shock, especially on the presynaptic terminals of the siphon sensory-gill motor synapses. This activity produces increased neurotransmitter release by the siphon sensory presynaptic terminals upon subsequent

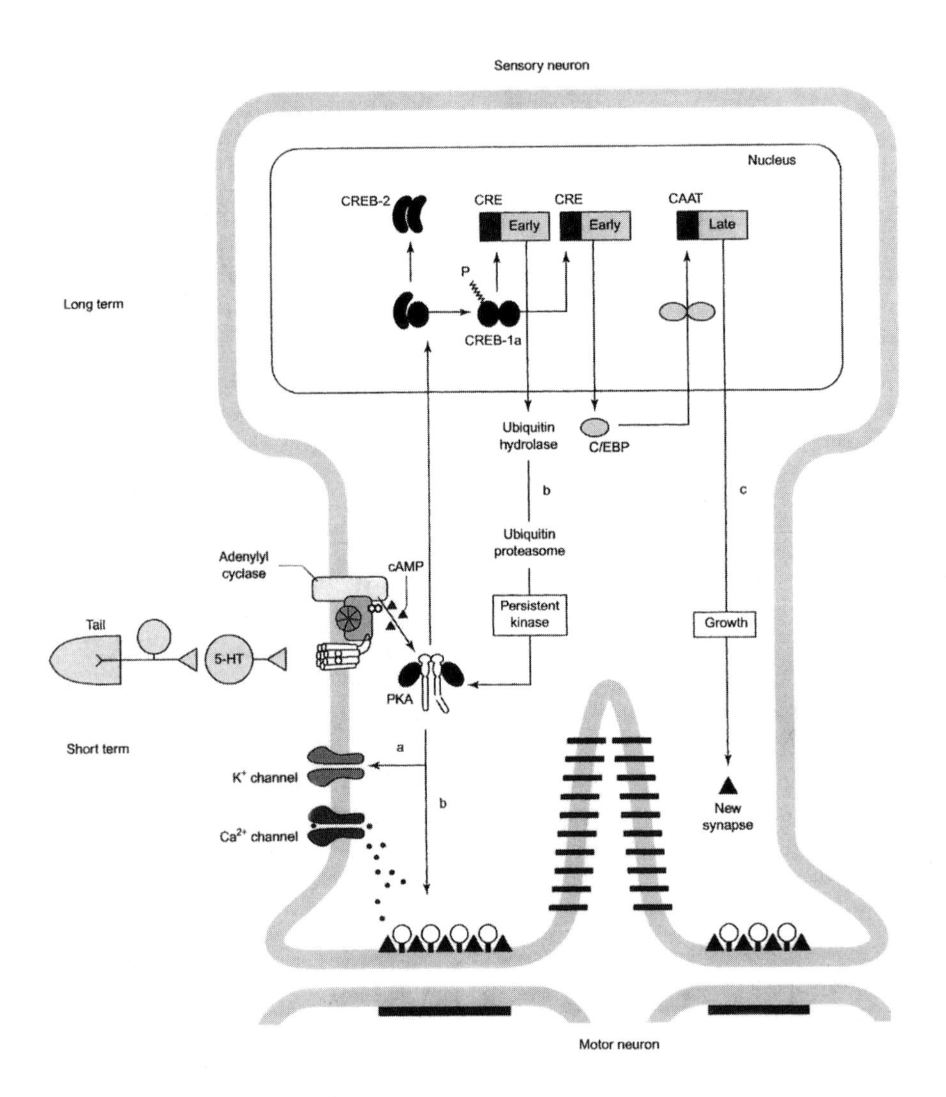

Figure 3.6. Molecular mechanisms of synaptic facilitation in *Aplysia* (sea hare). Molecular genetic effects producing long-term facilitation in presynaptic neurotransmitter release machinery were first discovered in this experimental preparation, but have since been found in *Drosophila*. See text for discussion. Reprinted from *Trends in Genetics*, **15**, "Genetics approaches to memory storage," 463-470, Copyright 1999, with permission form Elsevier Science.

siphon stimulation, and in turn heightened gill motor neuron activity and behavioral response (gill withdrawal). The intracellular molecular mechanisms of sensitization are exactly the ones involved in *Drosophila* olfactory conditioning (Figure 3.6; compare with Figure 3.5).

A nice methodological feature of the *Aplysia* monosynaptic siphon sensory-gill motor circuit is that its components can be extracted and reconstituted *in vitro* in dissociated cell culture. The modulatory interneurons driven by the (prior) tail shock release a catecholamine, serotonin (5-HT), as their neurontransmitter. Montarolo *et al.* (1986) showed that short-term facilitation in this monosynaptic preparation could be induced with a single puff (1micromole, or μM) of 5-HT directly onto the presynaptic terminal, and that long-term facilitation measurable for more than 24 hours could be induced by five puffs of 5-HT delivered over one-and-one-half hours. In both cases, facilitation was measured experimentally as a percentage increase in excitatory postsynaptic potential (EPSP) in the gill motor neuron. The 5-HT binds to metabotropic receptors in the siphon sensory presynaptic terminals, activating an intracellular G protein complex and priming adenylyl cyclase to generate more cAMP from ATP. The additional cAMP binds to the regulatory subunits of PKA molecules, freeing the catalytic subunits. In the short term (i.e., following a single puff of 5-HT), the catalytic PKA subunits phosphorylate K^+ channels in the presynaptic terminals and act directly at active zones. When the weak siphon stimulus occurs subsequent to these tail shock-induced short-term changes, the resulting action potentials in the siphon sensory neurons are broadened. More neurotransmitter is released, more binds to receptors on gill motor neurons, action potential rates increase in gill motor neurons, and the gill withdrawal response is dramatically stronger and quicker than under conditions of no sensitizing stimulus (Figure 3.6). Once again, all these molecular mechanisms are familiar.

Short-term classical conditioning in *Aplysia* is an elaboration of these same molecular mechanisms. The principal difference between sensitization and classical conditioning is the order of stimulus presentations. In sensitization the aversive stimulus comes first, followed by the neutral stimulus. In classical conditioning this order and timing is exactly reversed. A common assay for *Aplysia* conditioning also pairs a tactile siphon CS with a tail shock US. Optimal conditioning produces an even higher level of excitatory neurotransmitter release by siphon sensory neurons than sensitization produces. In the short term, this stimulus-specific increase results from the influx of Ca^{2+} into the siphon sensory presynaptic terminals by the action potentials generated by the *prior* siphon stimulus. The rest of the molecular story will be familiar. Ca^{2+} binds with intracellular calmodulin (CaM) to produce increased levels of Ca^{2+}-CaM complex. Additional Ca^{2+}-CaM molecules bind to additional adenylyl cyclase molecules, priming their efficiency for converting

ATP into cAMP. When the 5-HT released by the modulatory interneurons (driven by the now *subsequent* tail shock) binds to metabotropic receptors in the siphon sensory presynaptic terminals and activates the intracellular G protein complex, the primed adenylyl cyclase molecules convert far more ATP into cAMP. Increased cAMP yields increased free PKA catalytic subunits, which phosphorylate more K^+ channels in the siphon sensory presynaptic terminals and enhance neurotransmitter release. Increased neurotransmitter release upon subsequent CS presentations yields greater gill motor neuron response. And greater gill motor neuron activity generates a quicker and stronger gill withdrawal, even quicker and stronger than that produced by sensitization due to the primed adenylyl cyclase molecules generated by the now *prior* siphon stimulus (Figure 3.6).

But it is in the study of the molecular basis of the consolidation switch that the methodological advantages of the *Aplysia* experimental preparation really pay off. It is not only that the same mechanisms discovered in *Drosophila* olfactory conditioning have been verified directly and elaborated further in these more accessible neurons and circuits. Also, it is here that discoveries in invertebrates connect up closest with the molecular mechanisms of consolidation found in the wider variety of mammalian learning and memory.

Multiple spaced puffs of 5-HT to the presynaptic terminals *in vitro*, or multiple tail shocks (sensitization) or siphon touch-tail shock pairings (classical conditioning) *in vivo*, produce enough free PKA catalytic subunits via the cAMP second messenger pathway that these molecules translocate to the presynaptic neuron's nucleus. Notice once again that these procedures are laboratory equivalents of repetition, known through behavioral studies to produce memory consolidation. Using standard techniques from molecular biology, Bartsch *et al.* (1998) cloned an *Aplysia CREB1* gene and characterized its nucleotide sequence and the amino acid sequence of its predicted protein products. One of these products, the CREB1a polypeptide isoform, displayed 95% amino acid sequence homology to mammalian CREB proteins, meaning that 19 out of every 20 amino acids in the protein sequences were identical across these widely divergent species. Furthermore, the key phosphorylation consensus site in the *Aplysia* protein's phosphorylation (P) box, the site where freed PKA catalytic subunits induce their effects, is completely conserved between *Aplysia* CREB1a and mammalian CREB. *Every* amino acid is identical across the P box sequences.

To investigate the role of this protein product in *Aplysia* consolidation, Bartsch *et al.* (1998) used the monosynaptic siphon sensory-gill motor neuron circuit in dissociated cell culture. At two, four, or six hours before delivering either one or five spaced 5-HT puffs, they injected AsIV or AsIV/V, both antisense oligonucleotides, into the sensory neuron. These antibodies target specific nucleotide sequences corresponding to various exons in *CREB1a* messenger RNAs (mRNAs), interfering selectively with gene expression. They then measured percent change in excitatory postsynaptic potential (EPSP) amplitude in gill motor neurons, comparing activity in monosynaptic circuits containing injected versus uninjected neurons exposed to the same regimen of 5-HT puffs. Circuits containing the injected neurons exposed to a single 5-HT puff were normal compared to uninjected neurons when measured ten minutes later. This result indicates normal short-term facilitation (which does not require CREB-induced gene expression and protein synthesis). But long-term facilitation was abolished by anti-CREB1a antibodies when measured 24 hours later, indicating that the CREB1a protein is selectively necessary for inducing long-term facilitation in the *Aplysia* siphon sensory-gill motor circuit.

Bartsch *et al.* (1998) also extended the evidence that ApCREB1a is a transcriptional activator for memory consolidation. They showed that injecting recombinant CREB1a in combination with AsIV/V antibodies combined with five spaced 5-HT puffs rescued long-term facilitation, returning EPSP amplitude in gill motor neurons nearly to control levels when measured 24 hours later. They also showed that *Aplysia* CREB1a is the limiting factor in short-term to long-term synaptic facilitation. Injecting recombinant CREB1a protein into the presynaptic neuron coupled with a single 5-HT puff produced nearly as much facilitation (EPSP amplitude increase in gill motor neurons) as did five spaced 5-HT puffs to uninjected neurons. Finally, they showed that PKA-phosphorylated recombinant CREB1a is sufficient for producing long-term facilitation. Monosynaptic circuits containing injected neurons showed the same increase in EPSP amplitude 24 hours later without any 5-HT puffs compared to that induced by the standard five spaced puffs to uninjected neurons. This increase was completely abolished by co-injection of either an RNA synthesis inhibitor (actinomycin-D) or a protein synthesis inhibitor (anisomycin) into the sensory neuron. Standard molecular techniques also demonstrated that *Aplysia* CREB1a proteins are phosphorylated *in vivo* following 5-HT exposure.

The role of transcription repressor CREB2 in blocking consolidation of *Aplysia* learning and memory (sensitization and classical conditioning) has also been clarified experimentally. Bartsch *et al.* (1995) cloned ApCREB2, a transcription factor with a predicted amino acid sequence containing a basic-leucine zipper (bZIP) domain that interacts with *Aplysia* CCAAT enhancer binding protein (ApC/EBP).[26] It also contains consensus sequences for MAP kinase phosphorylation similar to those of human CREB2 and mouse ATF4 proteins. Standard molecular techniques show that ApCREB2 is expressed in sensory neurons, is a binding protein at numerous CRE sites, and represses ApCREB1a-mediated activation at CRE sites (all *in vivo*). When a single 5-

HT puff is paired with an injection of ApCREB2 antiserum to the sensory neuron *in vitro*, long-term facilitation measured 24 hours later was comparable to that following the standard five puff treatment to uninjected neurons. This effect is blocked by co-application of either a protein synthesis inhibitor (anisomycin) or an RNA synthesis inhibitor (actinomycin-D), indicating that the ApCREB2 antiserum-induced facilitation has the key properties of transcriptionally dependent long-term facilitation. Florescent micrographs of sensory neurons 24 hours after ApCREB2 antiserum injection and a single 5-HT puff display the characteristic new varicosities present in uninjected neurons following five puff treatment. However, ApCREB2 antiserum injection does not affect short-term facilitation measured one minute after a single 5-HT puff (compared to uninjected controls). Bartsch *et al.* (1995) conclude that ApCREB2 is a transcriptional repressor and that the ratio of ApCREB1a to ApCREB2 in presynaptic terminals determines the consolidation of short-term memory into long-term memory. This was exactly the mechanism suggested by Yin *et al.*'s (1994, 1995a) experiments with *Drosophila* single-gene mutants. A plausible mechanism of CREB2 inhibition is MAP kinase phosphorylation, which blocks its repressor activity. Exactly as in mammals (Chapter Two, section 4.2 above), PKA catalytic subunits don't bind directly to ApCREB2 molecules, but do influence MAP kinase levels in the neuron's nucleus.

The evidence from *Aplysia* is clear. Both phosphorylated CREB1a and CREB 2 are key transcriptional factors for consolidating sensitization and conditioning into their long-term forms. They are the next step forward in the cAMP-PKA intracellular path that influences synaptic change in invertebrates. But on which genes are these molecules acting? What protein products are being controlled by their fluctuating ratios? Again, the *Aplysia* experimental preparation proved advantageous. And the results will be familiar.

The role in memory consolidation of the immediate early gene *ubiquitin carboxyl-terminal hydrolase* (*uch*) and its protein product was first shown conclusively in *Aplysia*. Bergold *et al.* (1990) had shown that two-hour exposure to 5-HT lowers the concentration of PKA regulatory subunits but not that of catalytic subunits when both were measured 24 hours later, that cAMP is probably the second messenger mediating 5-HT intracellularly, and that this regulatory mechanism requires new protein synthesis. They conclude that this alteration in PKA subunit ratio somehow keeps the kinase in a persistently active state to produce the persistent phosphorylation seen in long-term facilitation. Hegde *et al.* (1993) showed further that degradation of an *Aplysia* PKA regulatory subunit (R) requires both ubiquitin protein and the proteasome complex it binds with, and that vertebrate PKA regulatory subunits (R_I and R_{II}) can also be degraded via the ubiquitin pathway. Building on these results, Hegde *et al.* (1997) showed that 5-HT, using cAMP as its second messenger, induces a neuron-specific *uch* gene in *Aplysia* (*Ap-uch*) whose protein product, ubiquitin hydrolase, has a similar amino acid sequence to a class of human uch. PKA phosphorylation-dependent ApCREB1a is the transcription activator for this *Ap-uch* immediate early gene. Its protein product, Ap-uch, has enzymatic activity dependent on the same residue as its human homologue and associates with the proteasome. Hegde *et al.* (1997) also clarified the role of Ap-uch in long-term synaptic facilitation by injecting sensory neurons of the *Aplysia* monosynaptic sensory-motor neuron circuit *in vitro* with antisense oligonucleotides that block the synthesis of Ap-uch. Circuits containing injected neurons treated with five spaced 5-HT puffs show no increase in synaptic facilitation when tested 24 hours later. However, the injections have no effect on short-term facilitation following a single 5-HT puff. The ubiquitin-enhanced proteasome degrades PKA R subunits, keeping the freed catalytic subunits in a persistently active state and subsequently enhancing neurotransmitter release for up to twelve hours following standard five puff spaced 5-HT treatment. Once again, readers of the detailed example in Chapter Two (sections 4 and 5) will be struck by a feeling of *déjà vu*.

Invertebrate-vertebrate similarities don't stop here. CCAAT enhancer binding protein (ApC/EBP) is a second transcription factor rapidly expressed in *Aplysia* sensory neurons following 5-HT treatment. Alberini *et al.* (1994) cloned the gene that encodes ApC/EBP. Its putative protein product is homologous in its amino acid sequence to rat C/EBP, especially in its bZIP domain. It contains a common consensus sequence within this domain for phosphorylation by both PKA and Ca^{2+}-calmodulin-dependent kinase II (CaMKII). The gene's regulatory region contains a CRE site, indicating that its expression could be activated or repressed by CREB proteins. The protein product ApC/EBP binds to numerous sites on a variety of early- and late-response genes further downstream. Alberini *et al.* (1994) also found that ApC/EBP is induced by 5-HT application in sensory neurons as an immediate early gene product, with cAMP serving as the second messenger. Long-term but not short-term synaptic facilitation is blocked selectively *in vitro* when sensory neurons are injected with an oligonucleotide that competes with ApC/EBP at CRE binding sites. Injections of ApC/EBP antisense RNA into the sensory neurons, which selectively inhibits ApC/EBP synthesis, likewise blocks long-term but not short-term facilitation. So do injections of antiserum BCA, a specific antibody against ApC/EBP. Alberini *et al.* (1994) also found that ApC/EBP needs to bind to its target regulatory elements for 9-12 hours to induce its long-term effects; this time frame is throughout the entire stabilization period for memory consolidation. These data, along with ApC/EBP's binding affinity for a number of late-response genes known to transcribe protein products necessary for presynaptic structural changes underlying lasting increases in neurotransmitter release rates, confirms its role as a transcription activator for protein synthesis-dependent long-term sensitization and conditioning. Furthermore, ApC/EBP is a downstream product of an immediate early gene for which CREB proteins are transcriptional effectors and repressors.

C/EBP completes the current model of memory consolidation in *Aplysia* (Figure 3.6 above).[27] Note that despite the experimental elaborations made possible by the *Aplysia* preparation, these mechanisms of nondeclarative memory consolidation (sensitization and classical conditioning) are identical to those of *Drosophila* olfactory conditioning. For fruit fries and sea slugs at least, multiple realization is nowhere to be found in the current molecular explanations of nondeclarative learning and memory. Perhaps that is not surprising. These species exhibit only the simplest forms of learning and memory in the simplest nervous systems. What is surprising—or "remarkable," as the *Drosophila* biologists claimed in the quote at the beginning of section 4 of this chapter—is that these same intracellular pathways, transposed into a postsynaptic key, underlie both E-LTP and its consolidation to L-LTP in forms of declarative memory specific to mammals (compare Figures 2.4, 2.6, 3.5, and 3.6). In the key molecular and genetic components, these homologies across species—from fruit flies and sea slugs to mammals—obtain down to the level of identical amino acid and nucleotide base pair sequences. In these quite specific respects, these "homologs" across species are much more precise and fine-grained than are standard "homologies" cited in evolutionary biology, e.g., human and whale forelimbs.[28] Here the molecular and gene compositions and the intracellular pathway interactions are shared across species. These shared features obtain despite vast differences in brain size, organization, site of principal effect (presynaptic or postsynaptic), behavioral repertoire, and even "cognitive logic" of the distinct types of memories being consolidated (declarative versus nondeclarative). Putnam's challenge has been answered empirically for one psychological kind, memory consolidation, that from the perspective of systems neuroscience seems obviously multiply realized. There is a "physical-chemical state," the cAMP-PKA-CREB molecular biological pathway, that uniquely realizes memory consolidation across biological classes, from insects to gastropods to mammals.

To understand the extent to which neurobiologists take these shared molecular mechanisms seriously, consider that in the Discussion section of one of their *Aplysia* publications, Bartsch *et al.* (1995) speculate that CREB2 inhibition might be a mechanism for the *human* "flashbulb memory" effect. "Flashbulb memory" occurs when information is stored after a single, usually emotionally charged occasion. This information can be retrieved for long periods after the event, sometimes for a lifetime. Personal memories from "when Kennedy was shot" or, for us younger folks, "when the Challenger exploded," are common examples. Bartsch *et al.* (1995) note that in mammals, the surprising and emotionally charged stimulus recruits activity in the amygdala and other catecholinergic modulatory systems. Perhaps these systems, via the familiar intracellular second messenger pathways induced by their modulatory neurotransmitters, temporarily relieve the repressive effects of CREB2 (and other transcriptional repressors), thereby "priming" the intracellular long-term facilitation and potentiation machinery to act in a fashion that normally requires multiple stimulus presentations. One experimental result reported in Bartsch *et al.* (1995) is pertinent to their speculation. When they blocked ApCREB2 activity, they induced the same level and persistence of long-term facilitation *in vitro* with a single 5-HT puff that required 5 spaced puffs in untreated neurons. One piece of evidence they cite for their speculation about human "flashbulb memory" based upon an *Aplysia in vitro* study is the shared intracellular pathways induced by modulatory neurotransmitters initiating CREB-related synaptic facilitation and potentiation in fruit flies, sea slugs, and mammals. Putnam's challenge has not only been answered scientifically; scientists even use this answer to suggest novel explanations and predictions about related *psychological* phenomena.

6 EVOLUTIONARY CONSERVATISM AT THE MOLECULAR LEVEL: THE EXPECTED SCOPE OF SHARED MOLECULAR MECHANISMS[29]

With the current models of memory consolidation in a variety of species now before us, along with an overview of experimental data supporting them and a consequence drawn against multiple realization, antireductionists might switch to a *burden of proof* argument. Molecular neuroscience suggests a unitary "switch" for memory consolidation across all its forms and biological instantiations, and that is surprising. But as Putnam himself emphasized more than thirty years ago (and we noted explicitly above), the "brain state theorist" must show this same result for *every* psychological kind. The multiple realization premise in the standard antipsychoneural reduction/identity argument requires only that the relation holds for *some* psychological kinds. Even if we restrict our concern to the empirical aspect of Putnam's challenge, doesn't the burden of proof still lie on the reductionist's shoulders? Surely molecular neuroscience can't yet discharge this burden of showing unitary molecular realizers for every psychological kind shared across species!

Of course it can't, at least not directly. Memory consolidation is one of molecular neuroscience's current parade cases. Similar results for other psychological kinds aren't yet on offer. But this case exemplifies a general principle of molecular evolution that we should expect to obtain in the discipline's future successes, namely, the slower rate of evolutionary change in the functionally important ("functionally constrained") regions of enzymes, proteins and genes. Such a principle of evolutionary conservation is implicit in the often-cited dictum, "evolution does not start from scratch," but instead builds upon genotypes or phenotypes already present in populations. It is also hinted at in the following quotes from noted scientists. Insect biologists Dubnau and Tully remark that

> In all systems studied, the cAMP signaling cascade has been identified as one of the major biochemical pathways involved in modulating both neuronal and behavioral plasticity. Molecular characterization of the [*Drosophila*] learning mutants *dnc* [*dunce*] and *rut* [*rutabaga*] offers a striking convergence of data with studies of learning in *A. californica* [*Aplysia*]. More recently, elucidation of the role of *CREB*-mediated transcription in long-term memory in flies, LTP and long-term memory in vertebrates, and long-term facilitation in *A. californica* suggest that CREB may constitute *a universally conserved molecular switch* for long-term memory. (1998, 438; my emphasis)

Even more directly to this point, Squire and Kandel remark:

> These several findings have provided a new set of insights into *the evolutionary conservatism underlying the molecular underpinnings of mental processes*. The simplest memory capabilities, and those that seem to have appeared earliest in evolution, seem to be nondeclarative memories related to survival, feeding, mating, defense, and escape. As a variety of additional types of nondeclarative memory and then declarative memory evolved, *the new memory processes retained not simply a set of genes and proteins, but entire signaling pathways and programs for switching on and stabilizing synapse connections*. Moreover, these common mechanisms have also been *conserved through the evolutionary history of species*: they are found in both simple invertebrates such as *Drosophila* and *Aplysia* and complex mammals such as mice. (1999, 155; my emphases)

While suggestive, however, these remarks are vague. Happily, developments in molecular evolution can give them real scientific substance. The functional regions of molecules participating in particular types of intracellular biochemical pathways are the slowest to change across species sharing the ancestor that first possessed them. From well-established general principles of molecular evolution, finding the same molecular mechanisms at work in memory consolidation across present species turns out not to be so "remarkable." Nor will it be surprising to find similarly shared molecular mechanisms underlying other psychological processes that contribute to organisms' fitness.

Consider first a paradigm illustration of the general principle.[30] The insulin protein, a hormone that transports glucose into cells, is formed by removal of a central region (the C protein) from the proinsulin molecule. The two remaining ends of the proinsulin molecule then bond to form insulin. Comparing the amino acid sequences of the functional insulin molecule and the C protein across a variety of existing species shows that the C protein evolved six times more rapidly than the functional regions: 2.4×10^{-9} amino acid replacements per year in the C protein, or 24 per one hundred million years, compared with 0.4×10^{-9} in the insulin molecule, or 4 per one hundred million years. (This is in a polypeptide containing only 51 amino acids in its entire sequence; see Ridley 1998, 179-180). As diabetics diagnosed prior to the early 1990s know, before human insulin generated through recombinant DNA technology became readily available, injectable functional insulin came from cattle and pigs. Human ancestors diverged from cattle and pig ancestors long ago, and yet the functional insulin molecule remains virtually identical across these existing species. Insulin is but a single example of this empirically confirmed principle of extremely slow evolutionary change in functionally constrained regions of intracellular molecules.[31] A related principle holds for entire genes and proteins. The amino acid sequences of "housekeeping" proteins, especially ones that function in basic metabolic processes in all cell types (like insulin), evolve at much slower rates in all regions combined than do proteins with more specialized cellular functions or limited to specific cell types (see Ridley 1998, Table 7.1, 156).

These general principles of molecular evolution were already so secure empirically more than two decades ago that Kimura (1983, but going back originally to papers published in the late 1960s) used them as one argument for the *neutral theory of molecular evolution*. The neutral theory holds that most evolutionary changes at the molecular level—for example, changes in the amino acid sequence of a protein—are selectively neutral, conferring no increase in fitness. This view is in contrast to a selectionist theory of molecular evolution, which holds that evolutionary changes to molecular structures typically are advantageous, and hence a product of natural selection. Kimura (1983) insisted that the slower rate of evolutionary change in the functionally constrained regions of enzymes and proteins was best explained by the neutral theory. The neutral theory explains these facts by the higher probability that a change in the amino acid sequence in functionally less important regions will be neutral (neither deleterious nor advantageous). Any change in a functionally constrained region will almost certainly be deleterious to an organism's fitness, since it will probably affect the molecule's binding and interactive capacities. A change to a functionally unimportant region has a much greater chance of not affecting the molecule's binding capacities.[32]

Selectionists answered this challenge by applying a general argument that natural selection typically favors small changes in a trait over large changes. This general argument employs an analysis of adaptation usually attributed to R.A. Fisher. It is general in that it purports to apply to all types of traits, be they molecular or "macro." Consider a two-dimensional space with a quantitative measure of some trait x on the x-axis and a measure of its fitness on the y-axis (Figure 3.7). Assume that there is some value for x such that individuals of some species possessing x to that degree are optimally fit, with fitness declining away gradually from that value. This analysis creates a "hill" of fitness values for trait x. At any given time, the individuals of a well-adapted species will possess x to a degree that groups them somewhere around (but not at) x's peak fitness value. A small change in trait x, in terms of the length of a line segment along the x-axis from an individual's current location, will either move the individual slightly down the fitness hill (e.g., be deleterious) or slightly up (be advantageous). A large change in trait x will either move the individual further down the fitness hill (be very deleterious) *or over the top and down the other side*—leaving the individual lower on the fitness hill than it was before the large change in x. Large changes to a trait's measure in a well-adapted species are thus almost always deleterious (to the individuals undergoing them). Natural selection favors small changes. Applied to molecular evolution, a change to a protein's functionally unimportant regions might be a "small" change, and could (on rare occasions) contribute to increased fitness (move the individual up the fitness hill slightly) by "fine-tuning" the molecule's interactions. Natural selection will select these changes. A change to a molecule's functionally constrained regions, on the other hand, will typically be a "large" change. It will affect directly and significantly the protein's capacity to interact in intracellular processes. Thus it will either take the organism down the fitness hill directly, or over the top of the fitness hill and to a point down the other side lower than the point occupied prior to the change. In either case, natural selection will almost never select a change to a protein's functionally constrained regions. The chance of it selecting a "smaller" change to the amino acid sequence in a molecule's functionally unimportant regions is higher, and that explains the slower rate of evolutionary changes in molecules' functionally constrained regions.

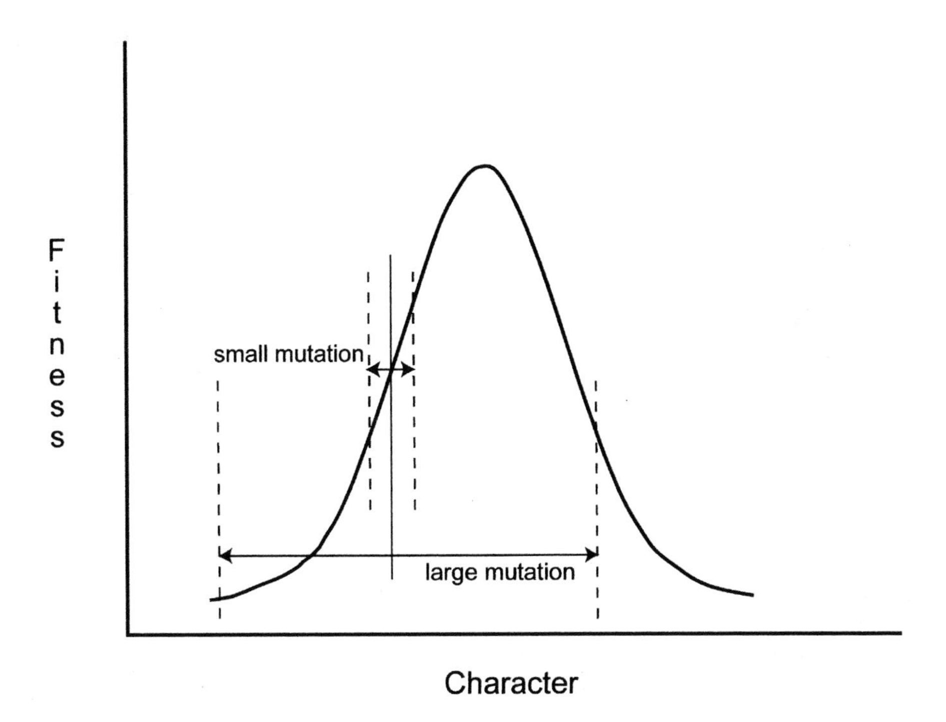

Figure 3.7. Fisher's model of adaptive evolution: the fitness hill. Fitness of a trait for an individual is graphed against degree to which the trait is possessed. The unbroken line represents the degree of the character possessed by individuals of a hypothetical well-adapted species. A mutation, that is, a change in the value of x, changes the bearer's fitness. Small changes to x have a higher probability of being positively adaptive (raising the bearer's fitness) than large changes. See text for full explanation. (See Ridley 1996, Figure B7.4, 182.)

The debate between neutral and selectionist theories of molecular evolution rages on to this day (Ridley 1998, chapter 7). Fortunately, we don't need to take a stand on it. The lesson we wish to draw from the empirical facts of molecular evolution holds regardless of which explanation is correct. The fact is, the rate of evolution for the functionally important regions of proteins involved in basic cellular metabolic processes is remarkably slow. This is the fact that both theories seek to explain. The amino acid sequences in these molecules and the intracellular processes they participate in tend to remain constant across species that share the ancestor that first possessed them.

How does this feature of molecular evolution support continual discoveries of shared cellular and molecular realizations of psychological kinds across species? Psychological kinds that contribute to survival and reproductive fitness must have some effects *on neuron activity*, on changing rates (or patterns) of action potentials in the neurons comprising various anatomical pathways and circuits. These commodities are neural currencies for information exchange within the nervous system and to the muscles that orchestrate behavior (Chapter Two, section 3.2 above). They are common to sensory transduction, neuronal communication, and output to motor effectors. They lie behind the basic "rate law" through which a given neural pathway represents information about distinct events (as differential rates of action potentials). Thus if psychological kinds have genuine causal effects on behavior (and are not just causally inefficacious epiphenomena), and so are factors that confer fitness on individuals, they must be realized in processes that affect action potential rates in individual neurons. In turn, a neuron's action potential rate depends directly on *its basic metabolic processes* governing intracellular energy exchanges and transformations, often through the effects of second messenger signaling cascades activated by neurotransmitters and hormones external to the neuron and impinging on its membrane receptors.[33] This means that the molecular mechanisms of any psychological kind that is causally efficacious and confers fitness on its possessors *must engage the functionally constrained regions of "housekeeping" proteins involved in basic cellular metabolism.* And as we just saw, these are the regions of biological molecules that *evolve the slowest.* Thus we should expect cellular and molecular neuroscience to find common molecular mechanisms across existing species that share psychological processes. At bottom, an active neuron is an active neuron, be it a fruit fly's, a sea slug's, or a mammal's. And these varying neural pathways and circuits that account for psychological differences across species are nothing but interconnected and interacting populations of membrane-bound molecular tricks for regulating ionic conductance, whose subcelllular and extracellular signaling components were selected for or drifted upon long before the large-scale variations evolved. Contemporary neuroscience is increasingly at the point where it can "link" processes occurring at these lower levels experimentally to behavior. It is this increasing behavioral evidence of specific psychological effects due to specific molecular manipulations within shared, evolutionarily-conserved intra- and interneuronal signaling and transmission pathways that make these accounts explanatory, and hence reductive, of psychological kinds.[34]

These general principles of molecular evolution inform another argument against multiple realization. The genes and proteins *common to dif-*

ferent cell types also evolve the slowest, and the cAMP signaling cascade and *CREB*-mediated transcription are certainly not unique to neurons. The cAMP signaling cascade activated by receptors coupled to intracellular G proteins is *the* classic second messenger system of molecular biology. It is active in cells of virtually every tissue type. In the paragraph that introduces the term "second messenger," the authors of a popular current undergraduate biology textbook write: "Cyclic adenosine monophosphate, or cAMP, is a well-studied second messenger that activates protein kinases in many different kinds of cells" (Purves *et al.* 1998, p. 850). Their diagram of a "typical" intracellular second messenger system (on the same page) illustrates a G protein-activated cAMP signaling pathway. Similarly CREB, along with cyclic adenosine monophospate response element modulator (CREM), constitutes *the* main class of gene regulatory activators of the cAMP signaling pathway. Gene regulatory programs induced by CREB control a variety of biological processes in a wide range of tissues besides neurons, including T cell development in the immune system, spermatogenesis, and the regulation of blood pressure through angiotensin (Haus-Seuffert and Meisterernst 2000). In the same textbook just cited, Purves *et al.* (1998) use cAMP-activated CREB-modulated gene transcription as their illustrative example of how surface receptors can trigger gene transcription in a variety of cell types. So the empirical facts underlying the arguments of the last two chapters are ones that academic biologists now teach to their freshmen majors! Not only are the molecular mechanisms of memory consolidation unified across biological species, but they also occur in cellular and developmental processes throughout the body.

From the molecular perspective, then, there is nothing inherently special about neurons. They are cells specialized to conduct electrochemical potentials down their lengths and affect this capacity in other neurons and (ultimately) muscle fibers. Neurons are neither "wonder tissue" nor a unique evolutionary "creation." They are a collection of molecular processes that are also at work in other biological tissues. The same molecular mechanisms underlying their collective functions (like memory consolidation) are at work in cells performing a variety of biological functions in other organ systems. Multiple realization in nervous tissue across species? Not even. When it comes to the underlying molecular mechanisms that drive everything biological, there isn't much "multiple realization" even across tissue types!

These discoveries give real substance to a prescient claim that Eric Kandel made more than two decades ago. Based on the barest suggestions about shared underlying molecular mechanisms available then, he invites us to "conceive of learning as ... a late stage of neuronal differentiation" (1979, p. 76): in other words, as one of the *developmental* processes that eventually individuate neurons![35] Since the beginnings of its "molecular revolution," leading neuroscientists have been aware of shared mechanisms across not only species but also tissue types and biological processes. Multiple realization in its broadest sense is inconsistent with readily-available molec-ular biological facts circa 2002. And ruthless reductionism is the vision guiding most cellular and molecular neuroscientists, along with most molecular biologists in general.

The proponent of empirical multiple realization might try one more counter-argument. Many philosophers are convinced that multiple realization of psychological kinds is true despite the fact that "unitary" *microphysical* accounts of the multiple realizers must exist. Few would dispute that the "distinct" physical events realizing one and the same psychological kind have commonalities at the level of atomic shells, electrons, and subatomic particles. The anti-reductionist's point is that unique realization at that level of description and explanation *is irrelevant* to psychology. Why not extend this reasoning up to the biochemical/molecular biological level? The anti-reductionist can grant that there will be "unitary" mechanisms there, shared across existing species. But why should that be any more interesting for the point at issue—*psychoneural* reductionism—than are the assumed commonalities at the microphysical level?

The problem with this reasoning is that assumed unitary mechanisms at the microphysical level are irrelevant to the multiple realization of psychological kinds. This is because at present microphysical descriptions have no explanatory relevance for behavior vis-à-vis psychology. Right now we can neither "explain behavior" nor manipulate it experimentally in properly specific fashion by mucking around directly at the level of microphysical posits. At present we cannot perform experimental manipulations by intervening directly on the microphysical level to generate specific behavioral effects for even the restricted sorts of behavior employed in controlled psychological studies, e.g., forced choices in a T maze or time spent freezing in a novel environment. We can't (now) explain psychological generalizations at the microphysical level, either.[36] But after nearly two decades of molecular neuroscience, we *can* explain quantitative behavioral data at the level of biochemical pathways and intracellular molecular mechanisms (in conjunction with facts about cellular constitution and anatomical circuits and pathways). And we can interfere *directly* with these molecular mechanisms in controlled experiments that yield *specific* behavioral data. We saw in the previous chapter how to explain quantitative data about the two key behavioral features of memory consolidation using the molecular mechanisms of LTP, the role of stimulus repetition and the effects of retrograde interference. Unlike microphysics, molecular neuroscience has explicit explanatory power for behavioral data, here and now. Unitary mechanisms already found at this level thus count legitimately against claims of multiple realization in a way that speculative future reductions to micro-physics do not.

Still, there is no reason for ruthless psychophysical reductionists to rest on the laurels of current cellular and molecular neuroscience. Biochemistry has yet to solve the "folding problem" of amino acid strings for proteins' tertiary structures, but advances on specific instances are being made. And once we have a grasp of that, it is likely that explanations of what current molecular biology has to assume (without explanation) will come forth. At some time in my future professional career, will I get to write a book for philosophers on the comprehensive *biochemistry* of behavior, cognition, and consciousness, with molecular neuroscience then an "essential heuristic"? The *really* ruthless reductionist in me hopes so!

7 CONSEQUENCES FOR CURRENT PHILOSOPHY OF MIND

It is now time for *scientifically inspired* philosophers to give up the multiple realization argument. Molecular neuroscience, the core of the discipline for nearly two decades, is showing that shared molecular mechanisms, conserved evolutionarily across present-day species, realize shared psychological features and processes. Experimental evidence is now clear for the "consolidation switch" from short-term memory to long-term memory, and this case reflects general principles of molecular evolution we should expect to discover for all molecular mechanisms affecting fitness. There should also be no shrinking from the consequences of rejecting multiple realization. With it goes the strongest—perhaps the only—*empirical* argument against psychoneural reduction. This leaves nonreductive physicalism, the most popular solution to the mind-body problem among current philosophical orthodoxy, without empirical support.

How might a *scientifically inspired* nonreductive physicalist respond? One possibility is to dig into current cellular and molecular neuroscience and find empirical evidence for multiple realization.[37] We've seen that this approach cuts against general principles of evolutionary conservatism at the molecular level. It will also require serious argument for why we should identify the psychological features across species with diverging molecular mechanisms.[38] This is an issue that few proponents of multiple realization have felt compelled to address, although at least one critic has raised it explicitly (Zangwell 1992). But at least nonreductive physicalists adopting this strategy could still count themselves as *scientifically inspired*. The other alternative is to give up on scientific inspiration and count oneself among the anti-empiricist metaphysicians of current philosophy. This is a classification that many philosophers of mind try hard to avoid. However, relying on an image of neuroscientific practice and results dating back to the 1970s (prior to the discipline's "molecular revolution") or on "possible world" intuitions and science fictional scenarios to defend one's key premises (e.g., multiple realization) are hallmarks of "armchair" philosophy. A choice is beginning to loom for philosophers of mind: either state-of-the-art neuroscience, which is ruthlessly reductionistic, or anti-empirical armchair metaphysics. Middle ground is disappearing quickly with the advancement of cellular and molecular neuroscience.

NOTES

[1] The other is explaining consciousness. We'll address it in Chapter Four.

[2] Heil and Mele (1993) remains a good introduction to these issues and the philosophical exotica they have spawned. Trent Jerde pointed out to me that the cognitive neuroscientific empirical literature also contains work on "downward causation." He cites Pardo, Pardo, and Raichle (1993) as an important contribution, as they study how self-control of mental states can direct neural responses. Jerde admits that a regress is looming. What about the neuronal processes constituting (or at least causally affecting) the self that is directing these responses? However, these empirical studies might hold promise for fruitful philosophical reflection, especially in light of the arcane concepts that have dominated purely philosophical discussions of mental causation.

[3] Horgan presents the claims to follow in a slightly different order. I present them in this order to emphasize the conditional nature of claims 2 and 3 (in my numbering), with 4 asserting the antecedent of the implied conditionals.

[4] We'll investigate multiple realization in great detail in sections 4-6 of this chapter.

[5] Horgan (2001) articulates and defends a version of causal compatibilism by applying David Lewis's ([1973] 1983) observations about implicit, contextually variable discourse parameters to concepts like 'cause' and 'causal explanation.'

[6] *Of course* there remains a "psychology of memory/" Consolidation is not the only important feature of memory, just the one that first yielded to "ruthless" cellular/molecular reduction.

[7] It seems reasonable to take "results published in *Cell*, *Neuron*, or similarly influential and respected mainstream neuroscience journals" as an adequate condition on "experimentally verified." If you doubt this, pick up an issue yourself.

[8] The "where to look" role is prominent in functional neuroimaging, using techniques that are prevalent in current cognitive neuroscience. I'll discuss this point in the next section of this chapter.

[9] It is worth noting that more than one century ago William James (1890) noticed similar sequential features in our "streams" of conscious experience. Given the apparent importance of frontal cortex in the neurobiology of consciousness, by discovering the cellular mechanisms of sequential features of cognition, we might also be drawing a bead on the ones underlying sequential features of consciousness, too. Bickle *et al* (2001) characterize these specific

sequential features and emphasize these similarities across cognitive processes and our "Jamesian" conscious streams.

[10] I describe the "dorsal" and "ventral" visual streams in some detail in Chapter Four, section 3 below. See especially Figures 4.2 and 4.3. That level of detail is not important for the current discussion.

[11] For FEF location, see Figure 4.1 in Chapter Four below. For a good "textbook" overview of the primate oculomotor system, see Goldberg *et al.* (1991).

[12] I am skipping over many scientific details here because they have already appeared in print. See Bickle *et al.* (2000).

[13] For the anatomical location of DLPFC, see Figure 4.1 in Chapter Four below. Sections 1 and 2 of that chapter contain a detailed discussion of Goldman-Rakic's and her colleagues work. For current purposes, this paragraph will suffice.

[14] In our complete model, we also developed a "Return to Fixation" mechanism that breaks off execution in the middle of a multiple-step saccade sequence and computes the dimensions of a saccade back to the original fixation point (Bernstein *et al.* 2000). We derived components of this additional feature directly from single-cell electrophysiology of "suppression site" neurons in FEFs (Burman and Bruce 1997) and structural MRI and neuropsychological assessment of two patients with anterior cingulate cortex (ACC) lesions encompassing the "cingulate eye fields" (Gaymard *et al.* 1998). Since we are just beginning to explore this component of our biological model with fMRI in behaving humans, I won't discuss it here in any detail.

[15] The next seven paragraphs describe the experimental task, fMRI data collection procedures, and preliminary data analysis first reported in Bickle *et al.* (2001). Since that report, we have fully processed and analyzed the preliminary data sets, so results reported here are new. I include some technical details to illustrate the complexity of even a quite simple functional neuroimaging task. A scientific manuscript describing the methods and results is currently in preparation. This project is completely collaborative, so the next seven paragraphs should be considered co-authored by Malcolm Avison, Vince Schmithorst, Anthony Landreth, and Scott Holland. Please note that co-authorship (and their full agreement with my arguments) does not extend to the final subsection of this chapter! I also thank Kathleen Akins for helpful written comments on a paper length treatment of the philosophical and scientific arguments of this section.

[16] This timing was also designed to make the 4-step sequences very difficult, to begin probing activation in anterior cingulate cortex.

[17] Both strategies also identified the anterior cingulate cortex.

[18] The fit function is a second order polynomial $Y = A_0 + A_1X + A_2X^2$, where Ys are the pixel values and X is the data point (frame) number (1, 2, ..., 320). The linear and quadratic components subtracted away are $A_1X + A_2X^2$.

[19] And the anterior cingulate cortex.

[20] We are conducting a follow-up study to explore these different time courses of activation in these two prominent frontal working memory regions during saccade sequencing.

[21] We've now corrected this experimental flaw by introducing a fixation point during the control task!

[22] Thanks to Huib Looren de Jong and Maurice Schouten for emphasizing to me the importance of these worries.

[23] For nonphilosophers, Ludwig Wittgenstein ended his first major work, the *Tractatus Logico-Philosophicus*, with the following remark: "My propositions serve as elucidations in the following way: anyone who understands me eventually recognizes them as nonsensical, when he has used them—as steps—to climb up beyond them. (He must, so to speak, throw away the ladder after he has climbed up it." ([1919]1961). Please note that my appeal to Wittgenstein is metaphorical. As should be clear from the discussion in the text, I am not charging cognitive neuroscientists with literally asserting nonsense!

[24] There is always the issue about how much scientific detail to include in a book addressed to an interdisciplinary audience. More detail, comparable to the amount presented in the last chapter, is coming in this chapter. But the details are necessary, first and foremost to show how Putnam's empirical, scientific challenge has actually been met These details also speak to one of the general themes of this book, that impressive research and explanation of behavioral data is taking place in current cellular and molecular neuroscience, and philosophers of mind and cognitive science aren't aware of it. Thanks to Trent Jerde and John Symons for advising me to remind readers of the "bigger picture" that all the "gory details" aim to illuminate.

[25] In my (1998, chapter 5), I characterize this story as "combinatorial reduction." My account is designed explicitly to incorporate Hawkins and Kandel's evidence and arguments into a general theory of intertheoretic reduction.

[26] Short sequences of amino acids determine which of a handful of *DNA-binding domain motifs* a given transcription factor possesses. *Leucine zippers* consist of a stretch of amino acids with a leucine residue in every seventh position. DNA binding occurs at a stretch of positively charged residues adjacent to each zipper. CREB proteins and C/EBP possess the leucine zipper motif.

[27] However, Bartsch *et al.* (2000) have recently found another transcriptional activator in *Aplysia* neurons, Activating Factor (ApAF). ApAF is phosphorylated by PKA catalytic subunits and forms dimers with both ApCREB2 and ApC/EBP. These new results show that *ApAF* is a candidate memory enhancer gene further downstream from the CREB proteins. The (molecular) beat goes on ...

[28] Thanks to Huib Looren de Jong and Maurice Schouten for suggesting this standard example.

[29] This section was improved by discussions with Marica Bernstein and Robert Skipper.

[30] The next few paragraphs draw on Ridley (1998), chapter 7. This is a standard current textbook on evolutionary theory.

[31] Another textbook example is the heme region of the hemoglobin molecule. See Ridley, chapter 7, for detailed discussion.

[32] Though even here, most changes will be deleterious, since an amino acid replacement will typically affect the folded protein's tertiary structure. The neutral theory can thus explain the slow rate of evolutionary change even in proteins' functionally unimportant regions.

[33] Note that this fact holds for all current theories of neural coding, not just for frequency/rate coding. (See Chapter Two, section 3.2 above.) The early chapters of Kandel, Jessell, and Schwartz (2000) are a good introduction to the full range of basic metabolic processes in neurons. Those in Shepherd (1994) provide a more compact presentation.

[34] As discussed briefly in section 4 of this chapter, psychological differences across species are accounted for by different sequences and combinations of these cellular and molecular events—different sequences and combinations of the "cell biological alphabet"—made available by the more complex circuits and anatomies in "higher" cognitive species. However, I also repeat from that earlier discussion that psychological *differences* are not at issue in the multiple realization challenge; psychological *similarities* are.

[35] See Shepherd (1994) for a good (though increasingly dated) primer on the shared molecular and molecular-genetic mechanisms of synapse plasticity and neuron development.

[36] This nonexplanatory feature of current physics vis-à-vis psychology and behavior might be a reason why no one took seriously Paul Churchland's (1982) attempt to undercut the multiple realization argument by arguing that "reductive unity" for psychological kinds will ultimately be found in thermodynamics. Churchland not only failed to offer any real empirical evidence for this possibility, but even more importantly it is difficult to see (now) how thermodynamics could *explain* (in any genuine sense) concrete behavioral data—like, e.g., that from the Kandel lab's work with transgenic mice.

[37] Ken Aizawa suggested this in a commentary at the 2002 Southern Society for Philosophy and Psychology annual meeting. He claimed that, e.g., "protein kinase A" is defined functionally and multiply realized physically. I demurred, on empirical grounds. We agreed to leave the question open, pending further discussion.

[38] This work might be part of a general project that Jim Bogen suggests as a response to my arguments, that of developing a taxonomy of psychological kinds to see if anything systematic (and non-hand-waving) can be found that determines which kinds can and which cannot be "ruthlessly reduced." I would look forward to grappling with any proposed taxonomy.

CHAPTER FOUR
CONSCIOUSNESS

Consciousness is one psychological phenomenon thought by many to lie beyond the explanatory reach of cellular and molecular neuroscience. Recently some philosophers have appealed to features of consciousness to revive psycho-physical dualism (Jackson 1983; Nagel 1989). Others use them to urge "new mysterian" skepticism about our human cognitive capacity to solve the consciousness-brain problem (McGinn 1989). Some find in the "hard problem" of consciousness a call to revolutionize physics (Chalmers 1996; Penrose 1994). It is common for even those who are optimistic about neuroscience's explanatory potential to insist that explaining consciousness will require "whole brain" resources from cognitive neuroscience, such as sophisticated neuroimaging techniques, computational modeling in massively parallel neural networks, and dynamical/complex systems mathematics to analyze and interpret the results (Churchland 1995, Freeman 2000; Hardcastle 1995). Orthodoxy in the philosophy of mind and cognitive science holds that the techniques of traditional neurophysiology and their recent supplements from molecular biology won't be up to the task, even if cognitive neuroscience ultimately is. (Indeed, even the latter conditional remains deeply controversial.)

There are, however, alternative voices. Perceptual neurophysiologist William Newsome, for example, insists that "we have not yet begun to exhaust the usefulness" of traditional neurophysiology's "single unit approach," especially "the recent trend toward applying the single unit approach in behaving animals trained to perform simple cognitive tasks" (Newsome 1997, 57). He lists perceptual, attention, learning and memory, and motor planning tasks; each category has an obvious link with consciousness. More recently, Newsome has asserted that standard electrophysiological methods, updated with new technologies, provide the ultimate test for neuroscience's "most remarkable hypothesis":

> The most remarkable hypothesis of modern neuroscience is that the entirety of our *personal experience*—from our perception of the external world to our experience of internal thoughts—result solely from patterned electrical activity among the several billion neurons that comprise the central

> nervous system. Ultimately, the most stringent test of this hypothesis is to *create realistic experiences* and mental operations artificially, by directly activating known circuits of neurons in the brain in the absence of the external inputs that normally elicit such mental operations." (Liu and Newsome 2000, R598; my emphases)

Newsome's experimental work on this project is the current standard. He uses tungsten stimulating electrodes to elicit activity in tiny clusters of neurons. Microstimulation has been a standard technique in electrophysiology for decades, but technical developments and background knowledge now enable experimenters (like Newsome) to induce electrical activity directly in just a few highly specialized neurons, with remarkable behavioral results. (In section 5 of this Chapter, I'll present some of the scientific details and draw explicit consequences from them for the philosophy of consciousness.) And notice that in the above quote, Liu and Newsome's explanatory target is "personal experience." Consciousness in all its glory is at center stage for these reductionistic neuroscientists. Finally, notice that they speak of "creating realistic experiences" as the ultimate test of neuroscience's guiding hypothesis. This feature is in keeping with the theme developed in the last two chapters, namely, the need to manipulate directly at the cellular and molecular level to induce specific behavioral results as a condition on the cellular/molecular account's being deemed explanatory.

In fact, there is much work in current reductionistic neuroscience, particularly at the cellular level, that speaks to philosophical concerns about consciousness. Unfortunately, these results are not known to philosophers (or to many cognitive scientists), who presume to speak of neuroscience's "explanatory potential and/or limits." All these results constitute a progressive scientific approach aiming ultimately at explaining consciousness by explaining piecemeal its basic features. They amount to a convincing case against the orthodox idea that consciousness is beyond the pale of reductionistic neuroscience, or that reductionistic neuroscience has nothing to contribute to its scientific investigation. The purpose of this chapter is to make this case. Here I'll discuss:

1. experiments and results revealing the "working memory fields" of individual primate prefrontal neurons,
2. the effects of explicit attention on action potential profiles in individual visual neurons,

and

3. microstimulation studies on tiny clusters of visual and somatosensory neurons that induce phenomenological experiences in primates.

The last topic even hooks up with the clinical neurological literature on human patients undergoing brain surgery while awake. As before, the scientific details aren't easy, but the payoff of the trek is worth the effort—unless one is committed to a mysterian philosophy of consciousness come what may.[1]

1 PREFRONTAL NEURONS POSSESS WORKING MEMORY FIELDS

Recall from Chapter Two above that short-term memory retains information only temporarily and is very susceptible to distraction and interference. It is now common among cognitive psychologists to divide short-term memory into a number of distinct kinds, processes or capacities. One of these is *working memory*, which is usually thought to be a temporal extension of *immediate memory*. Working memory holds recently recalled or acquired information "on line" in service of ongoing cognitive tasks, including comprehension, reasoning, and problem solving (Baddeley 1994). Often-cited everyday examples include retaining a telephone number one was recently presented while dialing or waiting to dial, or composing and speaking a complicated sentence. These everyday tasks are nicely reflected in equally common metaphors for working memory, including "the mind's blackboard" or "global workspace." Like immediate memory, working memory has a limited capacity for the number of distinct items occupying it at any given time—psychologist George Miller's legendary "magic rule of seven [items], plus or minus two" is often cited for it—but a vast access to all types of memory items. Working memory also has a limited time scale: on the order of up to 20 seconds, but perhaps extendable to a couple minutes with active rehearsal, after which the items are forgotten or consolidated into longer term memory forms. Psychologist Alan Baddeley (1986) famously divides working memory into three components. A *central executive workspace* is supported by two "slave" systems: the *phonological loop* allows recycling or rehearsal of small bits of verbal information, while the *visuospatial sketch pad* stores images like faces or spatial layouts. The key idea is a temporary memory capacity, constantly updated for content as cognitive tasks and demands change that require readily available information. Distinct neural information processing systems appear to possess their own working memory regions (Goldman-Rakic 1996). Neuropsychologist Larry Squire and neurobiologist Eric Kandel suggest that "working memory actually consists of a relatively large number of temporary capacities, each a property of one of the brain's specialized information-processing systems" (1999, 85).

At least one prominent psychologist, Bernard Baars, stresses the continuity of working memory and consciousness. Baars writes: "All unified models of the mind have a small "working memory" that is closely associated with conscious processes. Working memory is that inner domain in which we rehearse telephone numbers to ourselves or, more interestingly, in which we carry on the narrative of our lives" (1997, 41). In his "theater model" of the human mind, working memory is analogous to the stage containing possible contents of current and upcoming conscious experiences, competing and cooperating for the "spotlight" of consciousness to shine upon them. The stage is limited in the number of actors that can be on it at any given time, but offers vast access to many different actors. Like consciousness itself, the procession of actors across the stage of working memory is typically serial and sequential, forming a stream pertinent to ongoing cognitive activities mostly occurring off the stage and in the audience of unconscious processes (Baars 1997, chapter 2). However, one need not share Baars's love of the theater metaphor to notice the close connection between working memory and conscious experience.[2] The common sense examples of working memory all involve first-person phenomenological consciousness. A friend barks out a telephone number to you: "6-9-8-6-8-7-4!" The telephone is inside the house. You rush in, rehearing the numerical sequence in conscious inner speech. The mobile phone is not on its pad. You keep up the constant stream of conscious inner chatter (and maybe even outer chatter, to which you also consciously attend). There's the phone! You consciously rehearse the number one more time as you dial. Working memory is inexorably tied up with ongoing conscious experience. Hence by unraveling the neural basis of working memory, including its cellular mechanisms, we are honing in on the neural basis of at least some of conscious experience.

The *delayed response* paradigm has proven to be an excellent experimental set-up for exploring the neural basis of working memory. In general, these tasks begin with a sequential presentation of the item or items to be recalled, followed by a delay period during which the item must be recalled, followed by recall or response performance cued to the initial item. For working memory studies the delay periods are short, typically only a few seconds. Target and test items can include shapes (e.g., faces), letters, words, or markers of spatial locations. Goldman-Rakic *et al.* (2000, Figure 50.1, 734) review some common designs of this general paradigm for working memory studies involving both human and nonhuman primates. One design that has proven especially illuminating for single-cell studies on nonhuman primates is the *oculomotor* delayed response task (ODR). In a version that tests working memory for spatial locations, the monkey fixates on a central point while a visual stimulus is flashed quickly in his periphery. The monkey maintains central fixation during a delay period after stimulus presentation, at the end of which the central fixation point is extinguished and the monkey must quickly respond by saccading (moving his eyes) to the spot of the stimulus target's remembered location. This task allows experimenters to record the monkey's exact direction of gaze throughout the task and also to know the exact retinotopic location of the target stimulus. The fixation requirement during the delay period insures comparable behavior on every trial and forces the monkey to rely upon mnemonic rather than postural cues.

Patricia Goldman-Rakic and her colleagues (Funahashi *et al.* 1989) used this ODR task on rhesus monkeys surgically prepared for chronic single-neuron recording while awake and alert. (This is the work I hinted at in Chapter Three, section 3.2 above.) To monitor eye position precisely, they implanted a search coil under the conjunctiva of one eye in each monkey and used a standard computerized monitoring technique. Fixation spot and visual stimuli are presented on a dark computer screen. The fixation point is a filled white circle that usually appears in the center of the screen, while the peripheral target stimuli are filled white squares. Each trial begins after a 5 second intertrial interval with the appearance of the fixation spot. After the monkey establishes fixation on the spot for 750 milliseconds (as monitored by the intraocular search coil technique), a visual cue appears for 500 milliseconds at one of eight peripheral locations (with location randomized over individual trials). Extinction of the visual cue is followed by a delay period from one-and-one-half to six seconds. The fixation spot remains illuminated during the visual cue and the delay periods; the monkey must maintain fixation on it throughout both periods or the trial is scrubbed and the monkey does not receive a reward for correct performance. Extinction of the fixation spot marks the end of the delay period. The monkey must respond within 500 milliseconds by saccading to the remembered location of the visual cue. A correct response requires a saccade that ends within a 6° diameter window surrounding the visual cue's actual location. The monkey receives a .2 ml drop of sweetened water as a reward for each correct response. To insure continued motivation, the monkey has been denied all liquids in its home cage for the previous twenty-four hours and works to satiety during each testing day (150-250 ml of liquid reward). Trained monkeys perform very well on this behavioral task, usually displaying over 90% correct responses for all visual cue locations across all delay periods (up to six seconds, the maximum delay period tested).

During trials, Funahashi *et al.* (1989) recorded from single cells using tungsten microelectrodes from 319 neurons in the prefrontal cortices of three rhesus monkeys. 288 of these neurons were located within or surrounding the caudal (back) half of the principal sulcus (PS) in dorsolateral prefrontal cortex (DLPFC) (Figure 4.1); these locations were confirmed by a later histological study. 170 neurons of these 288, or nearly 60%, displayed ODR task-related activity, in that their average discharge rate (spikes per second) during at least one task phase (fixation, visual cue, delay, or response) differed significantly from their intertrial interval average. Of these 170 ODR task-related PS neurons, 87 (30% of the total PS sample, 51% of task-related sample) had significant delay period activity differences. 69 of these 87 showed *directionally-selective* delay period activity differences that were statistically significant compared to intertrial activity rate only when the visual cue had been presented at one or two of the eight target locations. 50 of the 87 delay period neurons showed statistically significant *increases* in activity; 46 of these 50 showed directionally selective increases. In these 46 cells, activity rose quickly within 100 milliseconds after visual cue presentation (at the very beginning of the delay period) and ceased within 100-150 milliseconds after saccade initiation (response). Even those neurons showing significant delay period activity increases to more than one remembered target location typically had a preferred location that elicited maximal activity increase. Graphs of direction-selective individual PS neuron activity, with cue location on the x-axis and normalized measure of spiking frequency increase during the delay period on the y-axis, could be fit with to a Gaussian curve. For the 50 cells exhibiting directionally tuned delay period increases, the fit Gaussians tended to be narrowly tuned. These cells showed high activity increases for one cue location and actual decreases for all others. This makes their delay period activity computationally similar to the receptive field properties of visual neurons selective for some parameter of the external stimulus (e.g., line orientation, motion direction; see my discussion in section 3 of this chapter, especially Figure 4.4A).

Funahashi *et al.* (1989) also found that varying the length of the delay period (from 1.5 to 3 to 6 seconds) had no significant effect on PS neurons' activity. Directional selectivity remained constant and activity rate remained elevated throughout all delay period lengths they tested. In addition, for the handful of directional selective delay period PS neurons that were recorded from during one or more error trials, in which the monkey saccaded to a location different from where the visual cue had appeared, available data indicated that their responses were significantly depressed during the delay period on error trials compared to correct trials.

Based on these results, Goldman-Rakic and her colleagues attribute "memory fields" to these prefrontal PS neurons and hypothesize about the cellular mechanism of working memory:

> Each neuron with directional delay period activity had a "mnemonic" receptive field: only when the cue was presented in that field did the neuron show excitation or inhibition during the subsequent delay period. Moreover, directional

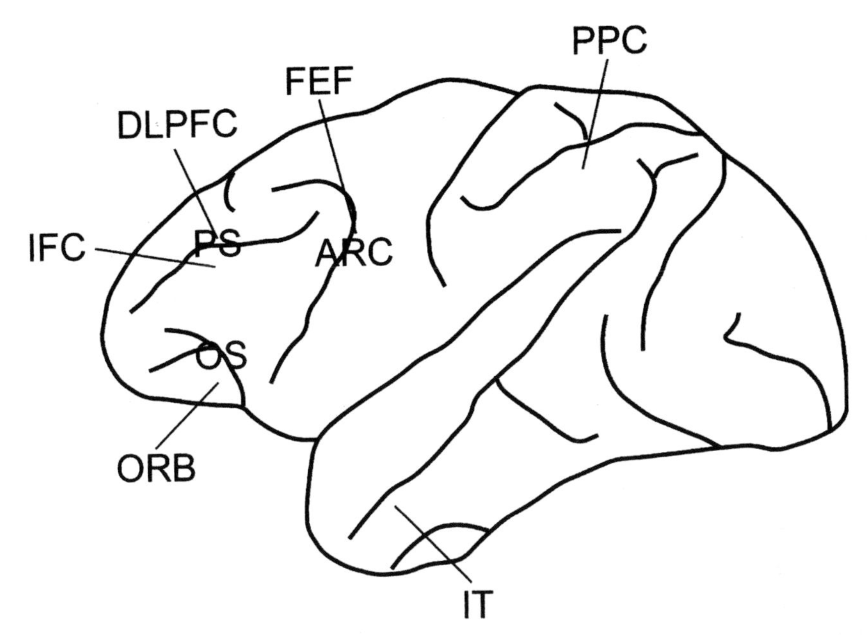

Figure 4.1. Schematic gross anatomy of primate prefrontal cortical regions (and other cortical regions) investigated by Goldman-Rakic and her colleagues. Labels: DLPFC, dorsolateral prefrontal cortex; FEF, frontal eye fields; ORB, lateral orbitofrontal cortex; PPC, posterior parietal cortex; IT, inferotemporal cortex; PS, principal sulcus; ARC, arcuate sulcus; OS, orbital sulcus; IFC, prefrontal inferior convexity. Figure created by Marica Bernstein, see O'Scalaidhe *et al.*, 1997, figure 3B, 1137.

> delay period activity expanded when the delay was lengthened and faltered on occasional trials when errors were made. Therefore we propose that this area of the visual field be termed the *memory field* of the neuron analogous to the receptive fields of visual neurons or the movement fields of oculomotor neurons. *Memory fields may be the cellular expression of a working memory process that allows mnemonic information to guide behavior.* (Funahashi *et al* 1989, 345; my emphases)

In the final sentence (italicized above), the authors explicitly cite Alan Baddeley as the cognitive-psychological authority on the kind of working memory they are proposing a cellular mechanism for. As I stressed earlier in this section, this is the kind of working memory tightly affiliated with conscious experience; and now we have a proposed cellular mechanism for a visual form of it, tied directly to single-cell data in behaving primates.

However, Funahashi *et al.* (1989) note two lacunae in their initial study. First, they can't rule out the possibility that delay period activity increases in PS neurons code for oculomotor dimensions of the upcoming saccade, rather than for the remembered spatial location of the visual cue. This alternative interpretation would block the link I'm stressing with a kind of working memory tied closely to conscious experience, since we are not consciously aware of impending oculomotor commands. All the experimental data they gathered is strictly consistent with this deflationary alternative interpretation.[3] Second, the mechanism by which a given PS neuron's memory field is constructed was not known, although the dense anatomical connections between DLPFC, posterior parietal regions known to process spatial visual information (being part of the well-known "dorsal," "where" or "how" visual stream) and some subcortical thalamic and limbic regions were suggestive (Figure 4.1 above and Figures 4.2 and 4.3 below).

The first of these lacunae was solved by a beautiful follow-up study (Funahashi *et al.* 1993). Goldman-Rakic and her colleagues compared responses of directional selective DLPFC neurons during delay periods on the ODR task of the (1989) study and an anti-saccade ODR task (AS-ODR). On the latter task, the monkey has to make a saccade in the direction exactly opposite the location of the visual cue after the extinction of the fixation signal indicates the end of the delay period. A filled white circle as fixation spot indicates an ODR trial; a white cross fixation cue indicates an AS-ODR trial. If delay period activity in directionally selective PS neurons codes for the oculomotor coordinates of the upcoming saccade, then it should be similar on ODR tasks with the visual target at one location—say, 90° from fixation spot, or vertically upwards—and on AS-ODR tasks with the visual target in the *opposite* location—e.g., 270° from fixation spot, or vertically downwards. Those two stimuli prompt correct saccades to the same location, namely 90° from fixation spot (vertically upwards). On the other hand, if delay period activity in directionally selective PS neurons codes for remembered spatial location of the visual cue, then a cell should have similar activity in the ODR and the AS-ODR tasks where the target cues are in the same location—say, 90° from fixation point (vertically upwards)—despite the different saccade dimensions required for correct performance (e.g., 90° vertical upwards in the ODR trials, 270° vertical downwards in the AS-ODR trials).

Funahashi *et al.* (1993) recorded from 108 prefrontal neurons in two rhesus monkeys during both ODR and AS-ODR trials. 51 of these neurons had statistically significant directionally selective delay period activity increases in the ODR trials. Of these 51 neurons, 44 could be localized histologically (after the behavioral and electrophysiological studies) to the PS or immediately adjacent cortex. 30 of these 44 neurons, or 68%, displayed delay period activity linked to visual cue location, not to oculomotor dimensions of upcoming saccade. (They responded in the second fashion described in the paragraph above.) The authors conclude from this result that "the preponderance of stimulus-dependent delay period activity in this area is strong evidence of a prefrontal specialization for ideational processing that does not rely on a motor code and is not mediated by motor signals" (Funahashi *et al.* 1993, 754).

These results on AS-ODR trials also strengthen the case for the hypothesis that activity in PS neurons with memory fields provides the cellular mechanism for the type of working memory tightly associated with conscious experience. The AS-ODR task requires the sort of mental inversion also required on such characteristically human cognitive tasks as the Stroop task or the Wisconsin card sorting task; performance on all three is seriously impaired in humans with prefrontal cortical damage.[4] Each requires that a habitual, sensory-driven response be suppressed, while an instruction-guided alternate response gets selected. As Funahashi *et al.* (1993) note, it is common in the human cognitive psychological literature to appeal to working memory as "a workspace for manipulation of symbolic representations" to explain performance on such "mental inversion" tasks. Once again they cite Alan Baddeley's work as their precedent for the type of working memory at issue. Even more clearly than on their earlier study (Funahashi *et al.* 1989), this follow-up study requires the kind of working memory closely tied to human first-person phenomenological consciousness. And again, they propose a cellular mechanism for it—activity during the delay period in PS neurons with directionally selective "working memory fields"—that emerges directly from single-cell primate neurophysiology *in vivo* during behavioral performance. Hence at least one psychological phenomenon closely tied to conscious experience yields to "the single-cell approach."

2 CONSTRUCTION AND MODULATION OF MEMORY FIELDS, FROM CIRCUIT CONNECTIVITIES TO RECEPTOR PROTEINS

What about the second lacuna in Funahashi *et al*'s (1989) original study, the unknown mechanisms of memory field construction in DLPFC neurons? The authors suggest that a circuit-anatomical account of principal sulcus (PS) neurons' afferents and efferents provide a hint. These neurons receive afferents from posterior parietal cortical neurons that comprise the end of the "dorsal" visual stream. Since the groundbreaking work of Ungerleider and Mishkin (1982) and of Goodale and Milner (1992), this pathway has been known to process information about the location of visual stimuli and the use of this information in coordinated movement. Previous studies suggested similar activity profiles between DLPFC and posterior parietal neurons during delayed response tasks, but Chafee and Goldman-Rakic (1998) were the first to compare these patterns directly. They recorded from the lateral intraparietal area (7ip) in the posterior parietal cortex and prefrontal cortical area 8a in the frontal eye fields under identical behavioral conditions, within the same hemispheres in two rhesus monkeys performing the ODR task (see Figure 4.1 above). They recorded from 252 posterior parietal and 235 prefrontal neurons whose firing rates changed significantly during at least one phase of the ODR task (compared with intertrial interval rate). The only differences they could find pertained to incidence and timing of firing. Parietal neurons responding to visual cue presentations fired earlier (and with a slightly higher frequency) than prefrontal neurons selective for that phase. A larger proportion of prefrontal neurons displayed delay period activity. But activity patterns in parietal and prefrontal neurons that displayed similar preferences for task phases were virtually identical. Both regions contained neurons with delay period activity and the spatial tuning properties of the neurons across these regions were the same (Chafee and Goldman-Rakic 1998, Figure 8). The authors note that these results suggest a principle of domain specificity by which prefrontal neurons share specific information through cortical networks of which they are a part. The fact that posterior parietal damage is *not* associated with spatial working memory deficits also suggests that these response similarities need not imply functional similarities. But they do provide an interesting hypothesis about the construction of working memory fields of prefrontal neurons, which are driven by similar neuronal activity in a class of neurons from which they receive direct inputs.

More recently, Goldman-Rakic and her colleagues have extended this reasoning with an intriguing suggestion about the evolutionary significance of prefrontal neuron activity:

> Organisms that primarily respond to sensory input and motor output may have developed the capacity to hold information on line by extending sensory responses to persist after the termination of sensory stimulation and thereby to flexibly instruct responses mediated by stored information, that is, information not available in the immediate stimulus environment. Ultimately, the elaboration of this process may have contributed to the human capability to behave independent of their immediate stimulus milieu and thereby

> to flexibly prepare for and think about future consequences of their actions. (2000, 740)

Expansion of prefrontal cortex is prominent in human evolution. With each experiment we are inching closer through working memory to full-blown human conscious experience.

This progress is crucial for the reductionist arguments in this book, because the remaining steps to report from Goldman-Rakic and her colleagues' work on the neural basis of working memory takes us increasingly *down* explanatory levels, to local cellular circuitries and ultimately to subcellular components. First, Funahashi *et al.* (1991) hypothesized that the different neurons that respond to different phases of the ODR task lie in distinct layers of a single cortical column in which all component cells have similar preferred remembered spatial locations. (This suggestion deepens the analogy with visual cortical organization.) These columns also include inhibitory interneurons that interact with pyramidal cells.[5] Goldman-Rakic and her colleagues (Wilson *et al* 1994) showed that these interneurons also display directional selectivity in the ODR task, often the inverse of pyramidal cells in their immediate proximity. So as an interneuron increases its activity rate, activity in surrounding pyramidal cells decreases. This suggests that local feedforward inhibition plays a role in constructing the memory fields of prefrontal pyramidal cells. Anterograde (forward-tracing) and retrograde (back-tracing) anatomical techniques have also revealed horizontal connections across cortical columns within specific prefrontal regions (Kritzer and Goldman-Rakic 1995). These results reveal "bands" that resemble iso-orientation columns in primary visual cortex. Based on these local connections, Goldman-Rakic (1995) proposed a model of working memory circuitry composed of clusters of pyramidal neurons with similarly tuned working memory fields, directly connected to each other (across cortical columns) by their excitatory axon collaterals. Inhibitory interneurons provide reciprocal interconnections between pyramidal neurons with opposite favored directional selectivity. These local circuit properties account for many of the single-cell results first reported in Funahashi *et al.* (1989).

The reductionist beat goes on. At the same time Goldman-Rakic was formulating her local model of memory field construction in prefrontal pyramidal cells, her lab was also descending down to the level of a specific receptor on these neurons for the neurotransmitter dopamine: the dopamine D1 receptor. Primate prefrontal cortex receives dense dopamine projections through the mesocortical pathway from the substantia nigra through the midbrain and forward to frontal cortex. Dopamine depletion in primate prefrontal cortex has been associated experimentally with behavioral deficits in a delayed spatial memory task, and prefrontal neuronal activity during a delayed response task has been augmented by dopamine applied iontophoretically (in tiny currents, to individual neurons, through special microelectrodes) and attenuated by a dopamine antagonist applied similarly (see Sawaguchi and Goldman-Rakic 1994 for references). To elaborate on the role of specific dopamine receptors in working memory, Sawaguchi and Goldman-Rakic (1994) employed the ODR behavioral task using delay periods from 1.5 to 6 seconds and a control task in which the visual cue remained illuminated during the delay period. On control trials, the saccade response is sensory-guided. Any behavioral impairment on ODR trials but not on control trials would thus indicate a specific working memory deficit, unaccompanied by a sensory or motor deficit.

Two rhesus monkeys were surgically prepared for precise eye location monitoring (using the standard intraocular search coil technique) and intracranial injections of drugs directly onto prefrontal regions while performing the behavioral tasks. Drugs injected during performances included two specific dopamine D1 receptor antagonists, SCH23390 and SCH 39166; a nonselective dopamine receptor antagonist, haloperidol; an inactive analogue of SCH23390, SCH23388; a selective 5-HT-2 (serotonin) receptor antagonist, ketanserin; a selective dopamine D2 receptor antagonist, sulpiride; a dopamine D2/D3 receptor antagonist, raclopride; and sterile saline. Six visual cue locations were used for both ODR and control trials. Drug injection sites were confirmed by a histological study after the experiment.

The various drugs were injected into 45 sites across DLPFCs of the two monkeys. Haloperidol injections, the nonspecific dopamine receptor antagonist, and injections of both specific D1 receptor antagonists, had significant effects on ODR performance at 22 sites, all clustered in a region in or immediately adjacent to the PS. Applications both decreased the accuracy of memory guided saccades and increased the latency of saccadic response (Sawaguchi and Goldman-Rakic 1994, Figures 3, 4, 5). Typically these deficits appeared 1-3 minutes after an injection, peaked after 20-40 minutes, and declined back to non-injection levels after 60-90 minutes. Injections of these drugs at DLPFC sites other than the PS had no effect on ODR performance. These deficits were always restricted to one or two target locations per effective injection site, a result in keeping with the earlier discovery of "working memory fields" of individual PS neurons. Memory-guided performance deficits were correlated directly with length of delay period. Longer delay periods (up to six seconds, the maximum delay period tested) were associated with increasingly inaccurate memory-guided saccades (to specific visual cues locations) and longer latencies (Sawaguchi and Goldman-Rakic 1994, Figure 6). The deficits were also dose-dependent. Larger injections (up to 60 micrograms of solution, the largest doses tested) were associated with increasingly inaccurate memory-guided saccades and longer latencies

(Sawaguchi and Goldman-Rakic 1994, Figure 7). On the other hand, these dopamine D1 receptor antagonists had no effects on accuracy or latencies of sensory-driven saccades on the control trials, at any delay length or dosage. Their effect was specifically on working memory, not the sensory or motor aspects of the ODR task. None of the other drugs—the selective dopamine D2 receptor antagonist, the D2/D3 receptor antagonist, the 5-HT-2 receptor antagonist, or saline—had any effects on accuracy or latency of memory-guided saccades on the ODR trials or sensory-guided saccades on the control trials, even when injected directly into the effective PS sites of the dopamine D1 receptor antagonists.

Based on these results, Sawaguchi and Goldman-Rakic remark that "it seems reasonable that the behavioral impairment induced by [D1] receptor blockade relates specifically to the working memory process now recognized to be a cardinal function of prefrontal circuits," especially ones involving PS neurons with directionally selective working memory fields. (1994, 522). Furthermore, their findings "provide evidence that the activation of the mesocortical dopamine system activates D1 receptors of the monkey PFC, *thereby modulating the mnemonic process associated with the PFC*" (1994, 524; my emphasis). And given that mesocortical dopamine fibers preferentially form synapses with pyramidal cells in primate prefrontal cortex, and that pyramidal cells containing a phosphoprotein associated with D1 receptors are primarily output neurons in cortical layer VI—projecting to thalamus and other cortical regions, ideally for communicating working memory information to ongoing and upcoming cognitive processing—Sawaguchi and Goldman-Rakic extend their explanation down an additional level. "It is, therefore, plausible that the mesocortical dopamine system specifically activates projection neurons located in the deep layers of PFC via postsynaptic D1 receptors on their basilar and/or apical dendrites in the upper strata" (1994, 525). This is increasingly a subcellular account of the mechanisms of working memory—of a type, I remind you, that is closely associated with conscious experience. Reductionism marches on!

The role of dopamine D1 receptors on memory field construction and modulation was further clarified—and reduced—by another follow-up study using the ODR task in primates outfitted surgically for iontophoretic application of drugs and single-cell recordings (Williams and Goldman-Rakic 1995). Their technique used a "quad-barreled" carbon-fiber microelectrode with one barrel outfitted for extracellular neuronal recording and the other three outfitted for drug delivery in miniscule quantities directly onto the neuron being recorded from. Williams and Goldman-Rakic discovered that SCH39166 delivery, the selective dopamine D1 receptor antagonist, at less than 30 nanoamperes (nA) injection current *accentuates* delay period activity increases in directionally selective PS neurons with memory fields during

ODR trials; but delivery at greater than 50 nA injection current virtually obliterated delay period activity, even after a stimulus in the neuron's favorite visual cue location. The drug had no effect on a neuron's activity during any phases of sensory-guided control trials. Raclopride, the dopamine D2/D3 receptor antagonist, had no significant effects on neuron activity during any periods of either ODR or control trials. Williams and Goldman-Rakic infer from these results that "the normal action of dopamine is to constrain neuronal activation during performance of a working memory task" (1995, 575). They then carry their hypothesized mechanism of memory field modulation down to the intracellular level: "A known mechanism of inhibition by D1 action is the attenuation of a slow inward sodium current which normally supports activation of the cell by excitatory inputs . . . Blocking the D1 receptor may simply disinhibit specific excitatory input to the same cells" (1995, 575).[6]

So far we've only seen these cellular and molecular mechanisms at work for the type of visual spatial working memory employed in ODR and AS-ODR tasks. Is there evidence for cells with memory fields and their molecular modulation in other types of working memory also affiliated closely with conscious experience? Goldman-Rakic and her colleagues (Wilson *et al.* 1993) found similar results in single-cell studies of neurons in the inferior prefrontal convexity below the PS (see Figure 4.1 above). They used a delayed response task keyed to nonspatial information about the color and form of a visual stimulus. This region was a promising candidate for such a study because it receives direct input from the inferotemporal cortex, the end of the "ventral" visual stream involved in object identification (based on such features as color and form). Lesions in this prefrontal area have been linked to memory deficits for objects. Finally, sensory neurons in the prefrontal cortex have foveal receptive fields, the area of highest acuity on the retina for visual parameters that figure in object identification. Monkeys were surgically prepared for chronic single-cell recordings in this area while performing an ODR task that randomly interposed spatial or pattern memory trials. On pattern DR trials, a stimulus pattern appears in the screen center (after the monkey establishes fixation on the central spot). The pattern is extinguished and the delay period ensues. After the delay period, the monkey must make a rightward saccade response for one type of visual pattern and a leftward saccade response for another type. Spatial DR trials are likewise simplified to only two visual cue locations, one left and one right of the central fixation point. Thus the two trials require the same response, but differ in the type of working memory that guides them.

Wilson *et al* (1993) found 31 neurons in the prefrontal inferior convexity that displayed delay period activity. 24 of these responded selectively during the pattern DR trials and 6 responded during both spatial and pattern trials. Most of these neurons were selective to only one of the two patterns (i.e., their activity during the delay period of pattern DR trials did not increase when the other pattern was presented). Pictures of faces (both monkey and human) were particularly favored stimuli for these neurons. Wilson *et al.* (1993) found some that were responsive during the delay period of pattern DR trials only to specific faces, with no response after others. This last result is especially interesting for the point stressed just above. Activity profiles of prefrontal inferior convexity neurons share numerous features with those in the inferotemporal (IT) cortex. The prefrontal inferior convexity receives dense projections from IT cortex, suggesting a similar "circuit property" contribution to those neurons' memory fields. Wilson *et al.* write: "These connections presumably provide signals about the attributes of foveal visual stimuli on which prefrontal circuits operate" (1993, 1957). One difference is crucial, however. Activity in IT neurons, even to favored objects, declines with object familiarity. Wilson *et al.* (1993) observed no such decline in prefrontal inferior convexity neuronal responses. Such sustained activity even with object familiarity would be fitting for neurons that hold such information transiently on line for other cognitive processes when the visual stimulus is no longer present.

More recently, Goldman-Rakic and her colleagues (O'Scalaidhe *et al.* 1997) have found further evidence for the modular organization of prefrontal cortex in a single-cell study with faces presented as visual stimuli. Although there was no explicit memory component to their task, they continued recording from prefrontal neurons for a few seconds after stimulus offset. In a deliberate attempt to record from as wide a range of DLPFC and lateral orbital prefrontal sites as possible, they recorded from over 1700 sites in three rhesus monkeys trained only to maintain fixation while a foveal stimulus is presented and for a few seconds afterwards. They found 46 neurons whose activity increased at some point during or immediately after face presentation, and which gave no statistically significant response increases to any other type of patterned stimulus. 44 of these 46 face-selective neurons were located in the prefrontal inferior convexity or the immediately surrounding lateral orbitofrontal cortex. The other two were located in a region of the frontal eye fields. None were located along the principal sulcus (PS) or in the immediately surrounding superior frontal cortex, the predominant sites of neurons with *spatial* working memory fields. Hence all these face selective prefrontal neurons lie in regions that receive dense projections from IT cortex. Many of these neurons showed a tendency to begin firing immediately after the face stimulus extinguished. This activity lasted throughout the entire 2500 milliseconds until the next stimulus appeared. Two of the monkeys in this study had been trained (as part of another study) to perform ODR tasks; but the third had never been trained on any memory task. The authors conclude that "the capacity for face-selective persistent firing and delay-period activity does not depend on intention to make a response, but appears to reflect an intrinsic property of the neurons' responses to visual stimuli" (O'Scalaidhe *et al.* 1997, 1137).

Finally, Goldman-Rakic and her colleagues have combined all this detail—single-cell physiology, circuit connectivities, and subcellular receptor effects (including NMDA receptors)—into a computational model and computer simulation that examines "the synaptic mechanisms of elective persistent activity underlying spatial working memory in the prefrontal cortex" (Compte *et al.* 2000, 910). Their results include activity profiles from simulated pyramidal neurons that "reproduces the phenomenology" of the single-cell recordings in behaving primates engaged in ODR tasks (see their Figure 4). The model descends levels in its biological realism and comparable performance measures down to neurotransmitter receptor channel contributions to neuron oscillations during delay periods (see their Figure 6). Simulated interneurons also display movement fields like those predicted by the local circuit model based on experimental single-cell results (see their Figure 7). The network's working memory performance is even resistant to distractions (see their Figure 8) and yields a number of novel predictions testable by single-cell *in vivo* and neural tissue slice *in vitro* experiments (Compte *et al.* 2000, 922).

Collectively, this research is a first-rate collection of state-of-the-art neuroscience, pursued methodologically at multiple levels but always aimed at explanatory mechanisms at the lowest level that technology and theory permit pursuing at any given time. Single-cell neurophysiology sets the stage for this developing explanation of primate working memory capacities. The explicit tie between this kind of working memory and conscious experience, stressed in both cognitive and "folk" psychology, indicates that at least some features of consciousness are not beyond reductionistic neuroscience, as many philosophers assume them to be.

3 EXPLICIT ATTENTION AND ITS UNREMARKABLE EFFECTS ON INDIVIDUAL NEURON ACTIVITY

Although he is best known in philosophy for espousing eliminative materialism, Paul Churchland nevertheless advocates the reality of conscious experience and its status as a key explanatory goal of neuroscience. He writes: "Consciousness is at least a real and an important mental phenomenon, one that neuroscience must acknowledge as a prime target of its explanatory enterprise" (1995, 213). He lists "steerable attention" as one of its "salient dimensions," writing that "consciousness is something that can be directed or focused—on this topic instead of that, on these things rather than those, on one sensory pathway over another, even if one's external sensory perspective on the world is held constant (1995, 214)." Hence the discovery of neurobiological mechanisms for explicit selective attention and its effects will be another step forward toward explaining principal features of consciousness.

However, it should be noted explicitly that Churchland himself advocates a "connectionist-inspired" neurocomputational explanation of selective attention based on "recurrent" (feedback) projections from higher back down to lower layers of neuron-like processing units (1995, 215-226, especially 217-218). While his account shares some rough structural features with the neurophysiological details I'll present in this section, the level at which he pitches his "neurocomputational perspective"—activation vectors across populations of densely-connected neuron-like processing units—suggests that he too is pessimistic about explaining features of consciousness at the cell-biological level. "Connectionist AI" uses some distinctively unbiological principles and posits, and its level of abstract modeling has been eschewed by practicing computational *neuro*scientists at least since compartmental modeling became popular in the mid-1990s. In light of his recent writings, Churchland's attitude toward the ruthless reductionism espoused in this book is unclear.

In this section I'll begin with the neural effects of explicit conscious attention. For that is where the "single-cell approach" has paid remarkable dividends. But these results in turn guide us toward a mechanistic account of attention itself, specifically for visual attention (a robustly conscious form of the phenomenon!) but generalizable beyond that single modality. And philosophically, the payoff of this research is important. At the level of the individual neuron, explicit attention appears to elicit a very mundane effect, as mundane as that elicited by simply turning up the salience or contrast level of external stimuli. Hence the second-to-last intuition on which consciophiles might build a case—that consciousness, even if ultimately explainable in neural terms, is at least a very special kind of neural event and cause—is being dismantled by reductionistic neuroscience. (We'll examine consciophiles' last available intuition, the problem of qualia and subjective experience, in the final sections of this chapter.)

Psychologists have studied the behavioral effects of explicit conscious attention for decades. These studies have confirmed a handful of generalizations. Directing explicit conscious attention to a spatial location in anticipation of a stimulus there both decreases response time to stimulus onset and lowers the threshold of detection. A common experimental paradigm goes back to Michael Posner (1980). Subjects are cued to expect a stimulus at one location with a high probability, and with only very low probabilities at other locations. They are then tested for their capacity to detect stimuli of various intensities or for the speed of their responses to stimuli appearing at all these locations. Subjects are faster to respond and are capable of detecting less intense stimuli at cued rather than uncued locations, a phenomenon Posner (1980) refers to as "attentional benefit." Downing (1988) extended Posner's paradigm and showed that attention directed to spatial locations also affects sensitivity and bias for brightness, orientation, and form discriminations there. This sensitivity decreases with distance from the cued location. More recently, Rossi and Paradiso (1995) showed that preferential processing measured behaviorally by detection performance on near-threshold stimuli is also specific to features independent of location. Subjects are better able to detect peripheral gratings at specific orientations presented at near-threshold intensities when performing a central attention task involving stimuli oriented to that same degree. They obtained similar results on near-threshold peripheral stimuli at specific spatial frequencies.

John Maunsell and his colleagues have studied the effects of explicit selective attention on cortical representations of visual information. As he stated nearly a decade ago, such findings "are changing the way we view the visual cortex," from an extractor of sensory attributes encoded in retinal images to an active processor that selects a limited portion of the visual image for concentrated attention and reshapes it to accentuate current interests (1995, 764-765). Maunsell is also committed experimentally to the "single-cell approach," writing that while technologies continue to develop for investigating human functional brain organization, "animal models remain the only source of detailed information about how neurons encode visual information" (1995, 765). As far back as a decade ago, his lab perfected a technique to isolate the effects of explicit attention on individual neuron activity throughout all regions of the visual processing hierarchies.[7] They use rhesus monkeys, surgically prepared for chronic single-cell recordings while alert and awake, as their experimental preparation. As Goldman-Rakic and her colleagues are accomplishing for working memory, Maunsell's group is using "the single-cell approach" to reveal the neural basis of another phenomenon closely associated with consciousness. This approach, we are assured by philosophers of mind and many cognitive scientists, can't possibly help us "explain consciousness."

Maunsell and his colleagues use a delayed matching-to-sample task as their behavioral measure of attention. The monkey fixates on a central spot and is prompted by a cue to attend to one region of visual space. The monkey then depresses a button to indicate readiness and sample stimuli appear 500 milliseconds later at two locations. Their stimuli include orientation bars, moving spots, or color patches. One stimulus appears at the spot that the monkey was prompted to attend, the other appears somewhere else in the monkey's visual field. The samples appear for 500 milliseconds and extinguish, after which a delay period of 500 milliseconds ensues, followed by the appearance of test stimuli. The monkey's task is to indicate whether the test stimulus *at the cued location* exactly matches the sample stimulus presented there (e.g., bars at the same orientation, motion in the same direction, same color patch). He either releases the "readiness" button within 500 milliseconds to indicate match, or continues to depress the button for at least 750 milliseconds more to indicate no match. The monkey must maintain fixation on the central spot during the entire trial, with eye position measured precisely using the intraocular search coil technique. He is rewarded only if he maintains fixation on the central spot throughout the trial and correctly indicates whether sample and test stimuli match or don't at the cued location. Matches and nonmatches at the two locations are uncorrelated, so the monkey gains no advantage by attending to the wrong (uncued) location. Given the time constraints, this is a difficult attention task, yet trained monkeys regularly approach 90% correct trials.

While monkeys perform this attention task, Maunsell and his colleagues record activity in individual neurons throughout striate and extrastriate cortex using standard extracellular techniques and analysis. The region of cortex being recorded from determines the nature of sample and target stimuli employed and the receptive fields of individual neurons being recorded from determine the spatial location of the two sample and test stimuli. One stimulus is located directly within the receptive field of the visual neuron being recorded from (the region of visual space in which stimuli elicit activity above baseline action potential rate); the other stimulus is located in a non-overlapping region directly across the central fixation point from the neuron's receptive field. Experimenters can then compare activity in an individual neuron under conditions of attention to its receptive field (the "attended" mode, as measured by correct performance on the matching-to-sample task when prompted to attend to the location of the neuron's receptive field) and conditions of attention directed elsewhere (the "unattended" mode, as measured by correct performance when prompted to attend to the other location). By presenting the same visual stimuli to the monkey in both "attended mode" and "unattended mode" matching trials, any differences in neuronal responses across the two trials reflect attention ("behavioral state") effects on single neuron activity.

In a (1995) review, Maunsell points out that simple versions of this behavioral task and neurophysiological recording had demonstrated signifycant attention effects on the activity in neurons as far back in the visual processing hierarchy as extrastriate areas V4 and MT (see Figures 4.2 and 4.3). He gives data from one neuron in area V4 that responded to the same stimulus in its receptive field with roughly a 50% increase in action potential rate when the monkey was attending to that stimulus than when it attended to the other stimulus (1995, Figure 2). He also shows an effect in MST when the monkey was forced to attend to only one of two motion stimuli both occurring within the neuron's receptive field. The neuron's action potential rate increased to nearly 100 spikes/second when the attended motion stimulus moved in its preferred motion direction, but quickly declined to below baseline response levels when the attended stimulus suddenly changed direction (1995, Figure 3). Maunsell concludes that "the widespread state-dependent modulations revealed by these studies show that the overall pattern of activity in V4 and MST can change markedly depending on what aspect of the visual scene is the focus of attention" (1995, 767-768). He even argues that there are reasons to think that the studies available circa 1995 *underestimate* to full range of state-dependent modulations on neuronal responses. We should take the attention modulations on neuronal activity that had been found so far as marking the "lower limit" of state-dependent contributions.

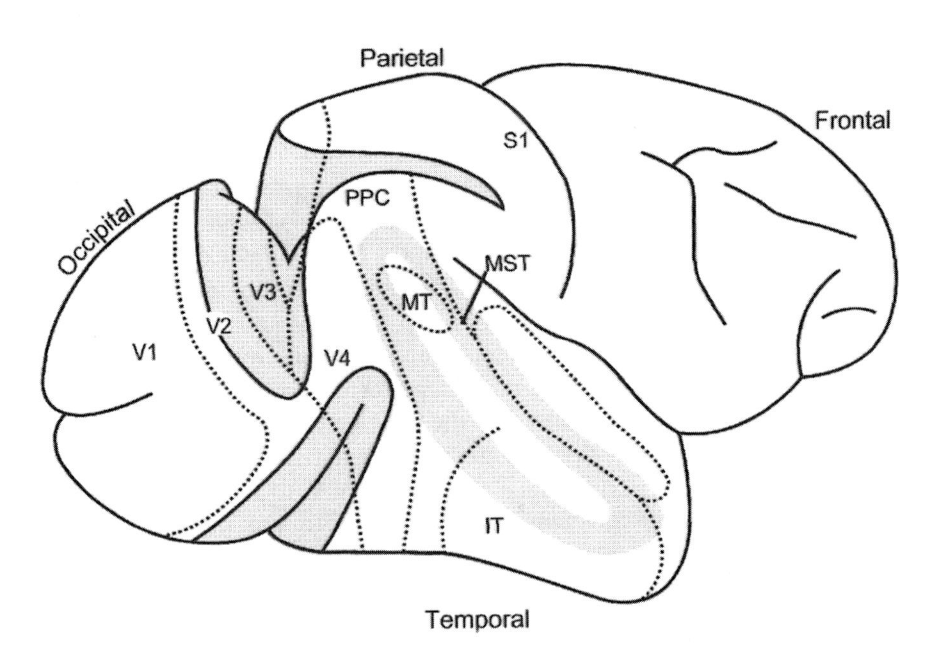

Figure 4.2. Schematic illustration of the gross anatomy of some sensory portions of primate cortex. Abbreviations: V1, primary visual cortex; V2, V3, V4, regions of extrastriate cortex; IT, inferotemporal cortex; MT, middle temporal cortex; MST, medial superior cortex; PPC, posterior parietal cortex; S1, primary somatosensory cortex. Reprinted (with additional labels) from *Fundamental Neuroscience*, M.Zigmond, F. Bloom, S. Landis, J. Roberts, and L. Squire (Eds.), 844, Copyright 1999, with permission from Elsevier Science.

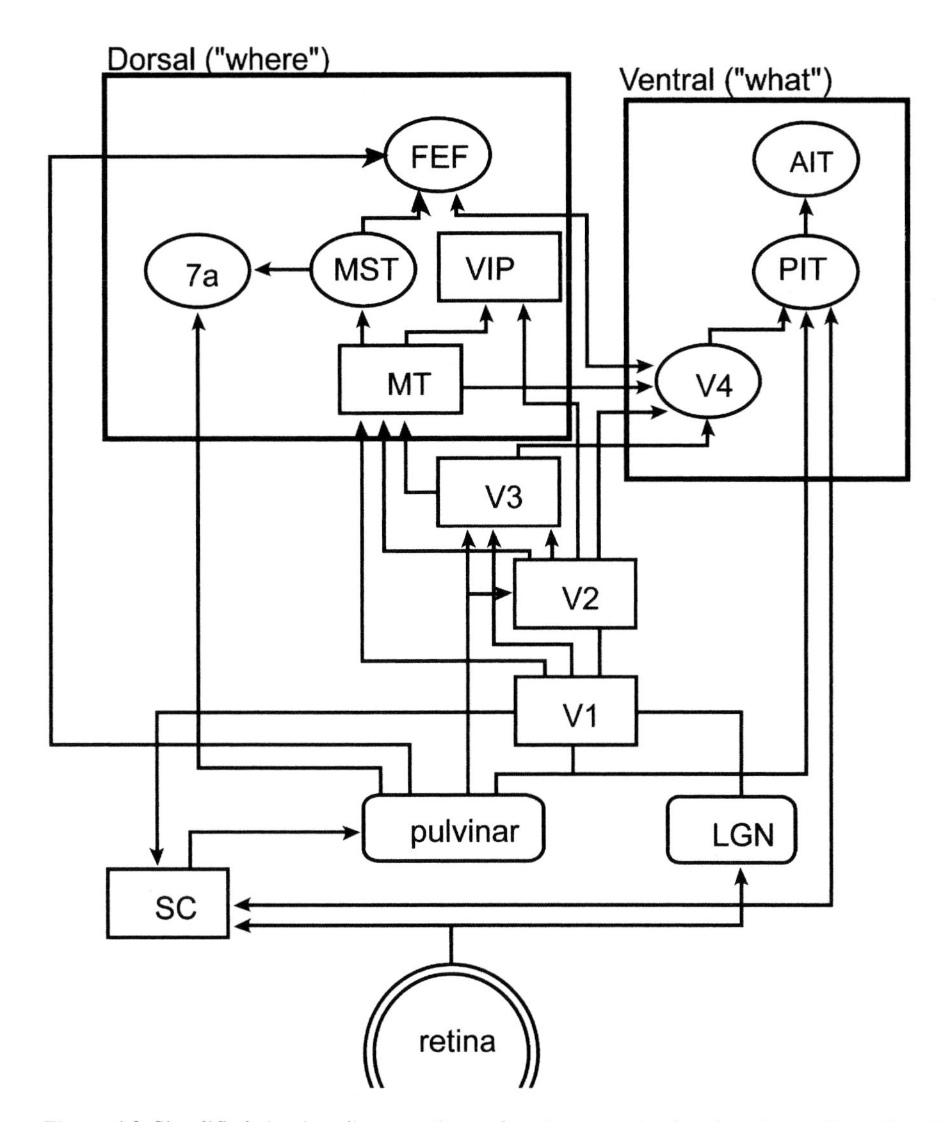

Figure 4.3. Simplified circuitry diagram of neural regions contained in the primate "dorsal" and "ventral" visual streams. Abbreviations are the same as in Figure 4.2, except: SC, (midbrain) superior colliculus; LGN, (thalamic) lateral geniculate nucleus; PIT, posterior inferior temporal cortex; AIT, anterior inferior temporal cortex; VIP, ventral intraparietal area; 7a, Brodmann's area 7a (in posterior parietal cortex); FEF, frontal eye fields. (Figure created by David Winterhalter.)

These findings of increased neuronal activity when a subject attends to a sensory neuron's receptive field correlate directly with psychological findings about the behavioral effects of explicit selective attention. But there are two ways that sensory neuronal activity modulated directly by attention state could produce faster reaction time and better discrimination at threshold stimulation. The simplest way is just to increase the neuron's action potential rate to all degrees of the stimulus parameter it responds to. Consider, for example, a visual neuron in area V4 that responds to stimulus orientation. The activity of such a neuron to orientation patterns in its receptive field fits a Gaussian curve with maximal activity (action potentials/second) elicited by one specific orientation, slightly less activity to related orientations, dropping down to baseline activity (or below) for opposing orientations (see Figure 4.4A). If attention modulates neuronal activity in this simple manner, the frequency of action potentials generated to a stimulus while attending to it will be greater at most or all degrees of stimulus orientation, as reflected in the increased height of its entire tuning curve under attentive as compared to unattentive conditions (see Figure 4.4B). The neuron's stimulus selectivity is unaltered, as reflected in the similar widths of its tuning curve under attended and unattended conditions. This effect is referred to as "multiplicative scaling" and it is a common neuronal effect. A similar effect in sensory neurons can be induced by simply increasing the salience of the external stimulus. The behavioral effects of attention, namely lower response thresholds and speed, would result because multiplicatively scaled neuronal responses have better signal-to-noise ratios and thus signal stimulus features more reliably. However, this mechanism would be deflationary for consciophiles because explicit conscious attention would then not be "specially" or "uniquely" realized neurally. It would be just another mechanism, albeit an internal/endogenous one, for "turning up the gain" on individual neuron activity. (And for that, there are numerous known neurobiological mechanisms.) Ho-hum for explicit conscious attention, from the brain's perspective.

On the other hand, explicit conscious attention could generate these behavioral effects by virtue of a much more robust and unique neuronal modulator. Perhaps it alters directly the *stimulus selectivity* of individual sensory neurons. Perhaps it sharpens effected neurons' tuning curves, increasing their activity to the most preferred stimulus degree and closely related ones, but dampening normal activity to those further removed (see Figure 4.4C). This would make neuronal activity in affected neurons signal more precisely the attributions of the attended visual stimulus. Sharper tuning curves provide more fine-grained representations of stimulus dimensions, which would improve both detection threshold and speed to attended locations and account for the behavioral data. Consciophiles also could be heartened if this is the cellular mechanism of attention modulation because increasing neurons' stimulus selectivity is not a common neurophysiological

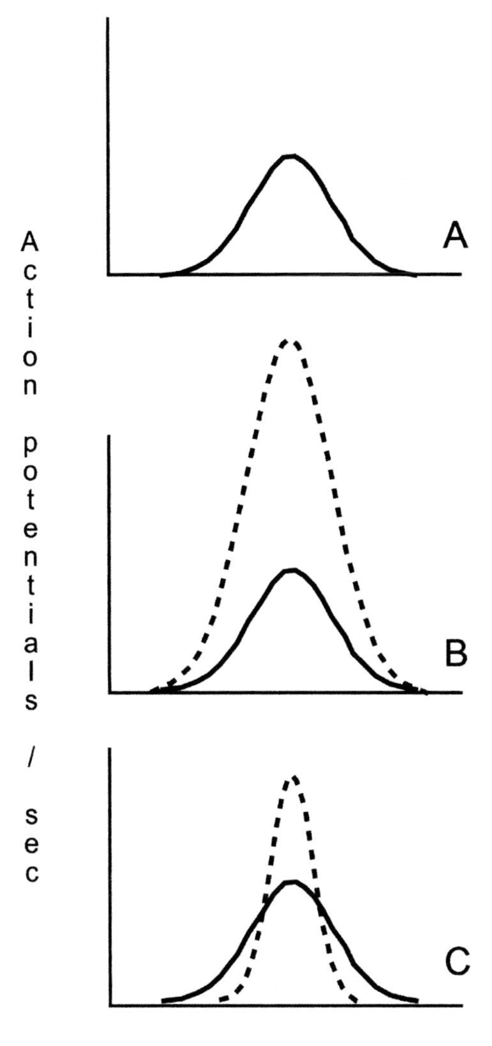

Figure 4.4. Tuning curves of an orientation-selective V4 neuron. A. Response under normal conditions fits a Gaussian curve, with maximal response (highest action potentials per second) to stimuli at a single orientation, near-maximal responses to orientations close to its most preferred stimulus, declining to responses blow baseline response levels to orientations increasingly different from its most preferred stimulus. B. "Multiplicative scaling," a general increase in the neuron's response to stimuli of any orientation, is one possible effect of explicit selective attention to the neuron's receptive field. C. A "sharpened" tuning curve is a second possible effect of explicit selective attention to the neuron's receptive field. Scaling and sharpening changes are exaggerated in B and C to better illustrate the possible effects. See text for further discussion and references to graphs of changes actually measured in V4 and V1 neurons under conditions of explicit selective attention to their receptive fields. (Figures created by Marica Bernstein.)

dynamic. It might even hearten mysterians, since known neurobiological mechanisms for this effect are not so readily apparent.

To test these competing explanations, Carrie McAdams and Maunsell (1999) have recently employed a version of the delayed matching-to-sample behavioral measure of attention described above with single-cell recordings in areas V4 and V1. Orientation bars always appear in the receptive field of the neuron being recorded from, while color patches appeared in the opposite location. The monkey is cued in the usual way about which location to attend to. "Attended mode" constitutes attention to the orientation bars in the neuron's receptive field while "unattended mode" constitutes attention to the color patch outside the neuron's receptive field. Monkeys must maintain fixation on the central spot throughout each trial and correctly indicate (by releasing or continuing to depress the button) whether the sample orientation bars match the test orientation bars (vertical or horizontal) when cued to "attended mode," or whether the sample color patch matches the test patch when cued to "unattended mode." By comparing individual neuron activity in "attended" and "unattended" modes to the same combination of visual presentations, McAdams and Maunsell isolate the modulatory effects of explicit selective attention on individual neuron activity. The time constraints on the behavioral task clearly elicit subjective conscious attention by human performers. By comparing a neuron's activity across the entire orientation tuning range under attended and unattended conditions, the debate over the nature of attention's modulatory effect can be resolved. Does explicit conscious attention elicit multiplicative scaling or a sharpened tuning curve?

Experimental results with over 200 orientation-selective V4 neurons and 121 V1 neurons clearly confirm the multiplicative scaling hypothesis (Figure 4.4B above; see McAdams and Maunsell 1999, Figures 2, 4, 5, 6, 7, and 10). For tuning curves of both individual neurons and averages across populations, the amplitude of attended responses (frequency of action potentials) compared to unattended responses to the same orientation stimulus was statistically significantly greater, for nearly all degrees of orientation. The only exceptions were orientation degrees that generated no response over baseline in the neuron. Explicit conscious attention to a visual neuron's receptive field enhances its action potential rate to virtually every degree of the stimulus parameter it responds to. However, the standard deviations of its tuning curves to the entire range of orientation degrees remain constant across "attended" and "unattended" modes. This means that the two tuning curves have nearly identical widths. Hence explicit conscious attention does not affect a neuron's stimulus selectivity. Finally, neurons' "attended" and "unattended" tuning curves had nearly identical asymptote values. This means that explicit conscious attention had no effect on neurons' responses to "unpreferred" orientation degrees. These results yield a decisive conclusion.

Directing explicit conscious attention to the location of a given sensory neuron's receptive field simply elicits multiplicative scaling. It only "turns up the gain" on the neuron's response to all degrees of the stimulus parameter, without sharpening its stimulus selectivity. Interestingly (and unlike some other studies that presumably used less sensitive measures and visual stimuli) McAdams and Maunsell (1999) found significant multiplicative scaling with attention all the way back to individual neurons in V1, the primary visual cortex, which is the first region of cortex to receive visual inputs in mammals via the lateral geniculate nucleus of the dorsal thalamus. However, they could not rule out that these small effects in V1 are caused by subtle position differences instead of attention modulation.

McAdams and Maunsell explicitly note that selective (conscious) attention therefore has the same effect on neurons as procedures as (metaphysically) mundane as increasing stimulus salience and contrast. They write: "The phenomenological similarity between the effects of attention and the effects of stimulus manipulations raises the possibility that attention involves neural mechanisms that are similar to those used in processing ascending signals from the retinas, and that cortical neurons treat retinal and attentional inputs equivalently" (1999, 439). Recently Charles Gilbert and his colleagues (Gilbert *et al.* 2000) have described extensive feedback projections from higher regions in the visual processing hierarchies and long-range horizontal connections within a given region that provide an explanation of endogenously-generated attention effects, as far back in the visual processing hierarchy as V1. This "plexus" of horizontal connections link neurons with widely separated receptive field locations, but with similar favored dimension degrees. In other words, cells that are maximally responsive to the same degree of a visual stimulus parameter, e.g., orientation direction, but to stimuli at very different locations in the visual field, are linked across cortical columns. This network is formed by the axons of cortical pyramidal neurons and the longest connections can span 5-6 mm, with multiple synapses from one end of the axon to the other. Functionally, this extends the area of visual space represented by the area of cortex from which individual neurons integrate input by an order of magnitude greater than the neurons' own receptive fields. "Contextual" influences, including attention, are likely to be mediated in part by these networks of horizontal connections and by feedback connections from higher cortical areas. Gilbert *et al.* even propose a cell-biological mechanism for these modulatory effects: "One possible mechanism underlying the attention effects is a gating or modulation *of the synaptic effects* of long range horizontal connections by feedback connections from higher cortical areas" (2000, 1224; my emphasis). This is what explicit selective conscious attention *is*, according to state-of-the-art current neuroscience.

Even more recently, McAdams and Maunsell (2000) have separated the correlates of spatial and feature-directed attention effects on individual V4 neurons, similar to the effects investigated behaviorally by Rossi and Paradiso (1995) described earlier in this section. Two rhesus monkeys surgically outfitted for eye position monitoring and single-cell recording while alert and awake were trained to perform two versions of the delayed matching-to-sample task. The first was a purely spatial attention task, where both the receptive field of the V4 cell being recorded from and the opposite location contained orientation stimuli. The second was a spatial and feature task, where the receptive field of the V4 cell being recorded from contained orientation sample and test stimuli, while the other location contained color sample and test stimuli. ("Attended mode," "unattended mode," and "correct trial" were defined as above.) By subtracting the average action potential frequency of a V4 neuron in "unattended mode" from "attended mode" and dividing by the average value in "unattended mode" for "space" and "space and feature" trials, any increase in the "space and feature" value for a given neuron would reflect increased attention modulation due to the feature component. For the entire population of 71 V4 neurons recorded from, McAdams and Maunsell (2000) found a 31% increase in activity during the spatial attention task, but a 54% increase in activity during the space and feature task. They conclude that "spatial attention and feature attention coexist in a relatively early stage of visual processing, cortical area V4," that "the same neuron can receive multiple types of attentional inputs," and that "directing attention to a stimulus feature might modulate the responses of neurons throughout the visual field," not just to the ones whose receptive fields contain that stimulus (2000, 1754). Notice that in their "space and feature" task, the color stimulus always appears at the location outside of the receptive field of the cell being recorded from.

What we have in the experimental work of Maunsell and his colleagues, coupled with that of Gilbert and his, is an explanation of explicit selective visual attention emerging from work at the "single-cell" level. Again, what some philosophers insist can't be done turns out to be a booming research program in mainstream current cellular neuroscience. But this research and its results carry an even more telling philosophical consequence. Some "consciophiles" have reconciled themselves with psycho-physicalism. They maintain this uneasy truce by holding out for some special, unique nature of consciousness's neural realization and effects. Even if consciousness is neural-cum-physical, it must be *a special type of neural-cum-physical event, cause, or effect*. But the upshot of the research discussed in this section denies even this much, for at least one prominent feature of consciousness. At the level of individual neurons, explicit conscious selective attention accomplishes nothing more than increasing external stimulus salience and contrast accomplish. It simply "turns up the gain" on neuron action potential frequency by way of endogenous activity across horizontal connections within a neural region and feedback projections from processing regions further upstream. This troubling consequence for consciophiles is by itself *philosophically* interesting. At the level of individual neurons, this feature of consciousness is *nothing special*. That this consequence was garnered by "single-cell neurophysiology" shows further the potential of "reductionistic" neuroscience, even for issues in the philosophy of consciousness.

4 SINGLE-CELL NEUROPHYSIOLOGY AND THE "HARD PROBLEM"

4.1 Chalmers on Easy versus Hard Problems of Consciousness

Now we can expect consciophiles to shed the kid gloves. So far we've seen how "single-cell" reductionistic neuroscience sheds light on two features of conscious experience: working memory and explicit selective attention. Sympathizers of David Chalmers, however, will point out that scientific explanations of *these features* of consciousness were never at issue or in serious doubt. Chalmers explicitly includes on his list of the "easy" problems of consciousness the following three: "the integration of information by a cognitive system; the ability of a system to access its own internal states; the focus of attention" (1995, 200-201). The first two clearly include the kind of working memory addressed by Goldman-Rakic and her colleagues; the third clearly includes the explanatory target of Maunsell, Gilbert, and their colleagues. Perhaps it is surprising that reductionistic neuroscience has already made such headway on these problems. Chalmers did predict eight years ago that "getting the details right will probably take a century or two of hard empirical work" (1995, 201). But hey, science is full of surprises of this sort. The key point is, we knew going in that those problems of consciousness were solvable by science.

This contrasts with "the really hard problem of consciousness . . . the problem of *experience*. When we think and perceive, there is a whir of information processing, but there is also a subjective aspect. ... This subjective aspect is experience. ... What unites all of these states is that there is something it is like to be in them. All of them are states of experience" (Chalmers 1995, 201). What separates the easy problems from the hard one is that subjective experience isn't exhausted by its functional properties and so can't be "reductionistically explained."[8] But reductive explanation is the broad method of the special sciences (sciences other than basic physics), including cognitive science, neuroscience, molecular biology, molecular genetics, and biochemistry. As these sciences stand, "they are *only* equipped to explain the performance of functions. When it comes to the hard problem, the standard approach has nothing to say" (1995, 204).

I leave it to readers to decide if what I will offer in the rest of this chapter is a step toward solving Chalmer's "hard problem." My guess is that committed mysterians about consciousness will deny that it is; but such denials will reveal the "pragmatic fruitlessness" of the hard problem, and the neo-Carnapian attitude I expressed in Chapter One bids us to ignore those. Over this section and the next three I will describe empirical progress that has been made in "inducing phenomenology" by manipulating tiny clusters of sensory neurons using a standard technique from traditional neurophysiology; suggest that this is a start toward explaining the neural basis of (sensory) qualia; draw out lessons from this research for an increasingly prominent position in the philosophy of consciousness, "phenomenal externalism"; and end by arguing that philosophers aren't the sole proprietors of "the problems of qualia and subjective experience"—that mainstream reductionistic neuroscientists have already staked a claim on them. I hope that readers not yet committed to the inexplicable nature of qualia and subjective experience will give this scientific evidence a fair reading.

There are two projects being pursued in this chapter. The positive project presents current scientific hypotheses and experimental results from cellular neuroscience about properties thought by many to be connected with conscious experience. The negative project is an implicit response to mysterians. The first is the project that most excites me, since it describes ongoing attempts to address features of consciousness scientifically. The empirical evidence shows that this scientific endeavor is progressing with demonstrated results. That this evidence can now be drawn from ruthlessly reductive cellular neuroscience is all the more intriguing, since few philosophers and cognitive scientists are even aware that these results exist. This final point is where the positive project of this chapter hooks up with the implicit jab at mysterians. Mystery thrives in ignorance; only now the ignorance is self-imposed because relevant empirical data is out there, in the cellular and molecular neuroscientific journals.[9]

4.2 Neuroscientific background: Wilder Penfield's pioneering use of cortical stimulation

Four decades ago, neurosurgeon Wilder Penfield published a comprehensive review of "experiential responses" elicited by electrical stimulation of the cortex in awake humans (Penfield and Perot 1963). He perfected this technique as part of a surgical procedure for treating otherwise intractable epilepsy. Since brains lack pain receptors, patients whose scalps, skulls and underlying connective tissue had been deadened with local anesthetics could comfortably remain conscious while surgeons ablated (removed) the site(s) of their seizure origins. Penfield's procedure was a clinical breakthrough. If electrical stimulation at a specific site evoked epileptic symptoms, this was evidence that the site is one of seizure origin. And by probing responses of conscious patients during announced and unannounced stimulations, the surgeon could explore the functional significance of tissue he was considering removing.

Penfield and Perot's comprehensive review reports that from 1938 to 1963, Penfield and his associates at the Montreal Neurological Institute performed 1,288 surgeries on 1,132 patients. 520 cases involved exposing and exploring the temporal lobe; 612 involved other neural regions. Electrical stimulation produced "experiential responses" in none of the latter 612 cases, while it did so in 40 of the former 520 cases (7.7%). Experiential responses were more complex than sensory experiences like whirring or buzzing sounds and color flashes or motor phenomena like involuntary limb movements. The latter were elicited routinely in many patients by electrical stimulation to appropriate sensory or motor cortical regions (Penfield and Perot 1963, 597). True experiential responses instead resembled the spontaneous "experiential hallucinations" and "dreamy states" characteristic of temporal lobe epileptic seizures. The ones surgeons induced by electrical stimulation were "sometimes extensive and elaborate, sometimes fragmentary," and often included "the sights and sounds and the accompanying emotions of a period of time, and the patient usually recognizes it spontaneously as coming from his past" (1963, 596). Auditory responses were most frequent, including a voice or voices, music, or other meaningful sounds. Experiences of music were surprisingly prominent. Visual responses were also frequent, often of a person or group of persons, a scene, or other recognizable objects. Auditory-experiential and visual-experiential responses sometimes occurred in combination, usually as scenes with appropriate sounds or a person or people singing or talking. In patients who commonly suffered from spontaneous experiential hallucinations during their seizures, electrically invoked experiential responses often resembled their spontaneous hallucinations. Experiential responses elicited from one site were often identical or similar to responses elicited from nearby sites.

Examples from the published transcripts of their forty case histories illustrate these features.[10] After removing the anterior tip of D.F.'s right temporal lobe (Case 5, 619-620), the surgeon stimulated a site on the cut surface of the superior and medial region of the first temporal convolution. On the second stimulation D.F. reported, "I hear some music." When the stimulation was repeated without warning, D.F. reported, "I hear music again. It is like the radio." She was unable to name the tune, but claimed it was familiar. Upon a later stimulation to this same site, D.F. reported, "I hear it." The electrode was kept in place and D.F. was asked to describe her experience. She hummed the tune. The operating room nurse named the tune and D.F. agreed with her judgment. The nurse agreed that D.F.'s humming captured the tune's proper timing and tempo. On further inquiry D.F. claimed that the experience was not that of "being made to think about" the tune, but that she "actually heard it."

E.C. had a history of "psychic precipitations" that always ended in seizure (Case 19, 632-633). Each of his attacks began after he saw someone grab an object from another person. The visual perception would produce a vivid memory of a time when he was thirteen, playing with his dog by grabbing a stick from the dog's mouth and throwing it. This association would confuse him and produce a seizure. During stimulation of a site just superior to the first temporal convolution in his left hemisphere, while E.C. was naming pictured objects, he reported, "There he is ... It was like a spell. He was doing that thing: grabbing something from somebody. It was somebody else doing the grabbing." When asked what he was grabbing, E.C. replied, "A stick, or something." When asked where he was, he replied "Up the street . . . That was like an attack, doing that thing." When the surgeon returned to stimulate this site again ten minutes later without warning, E.C. reported "There it is." The stimulating electrode was kept in place a short time longer and a major seizure ensued.

R.Re., a native South African, began suffering seizures eight years after recovering from a severe case of meningitis (Case 14, 628-629). During surgery a number of anatomically proximal sites were stimulated above the first temporal convolution in his left hemisphere, from the anterior tip to a point adjacent to the central sulcus. R.Re. offered the following sequence of reports: "Yes, something that someone has said ... Not here, in Johannesburg." "Yes, something that was said, also something that was said in Johannesburg, and it was said by somebody that had been put out." "Yes, I was hearing at Johannesburg, it came and went very clearly." "Yes, that same sort of sensation, somebody was speaking to me in Johannesburg." He mentioned that the speaking voice was different each time.

Based on their case histories and analysis, Penfield and Perot (1963) offered a number of clinical and neuropsychological conclusions. Some speak directly to the issue of *phenomenology induced by direct cortical electric stimulation*. For example, they write

> The conclusion is inescapable that some, if not all, of these evoked responses represent activation of a neural mechanism

> that keeps the record of current experience. There is activation too of the emotional tone or feeling that belonged to the original experience. The responses have that basic element of reference to the past that one associates with memory. But their vividness or wealth of detail and the sense of immediacy that goes with them serves to set them apart from the ordinary process of recollection which rarely displays such qualities. (1963, 679)

They emphasize the connection between evoked memories and the "stream of consciousness" during past experience:

> The true nature of [epileptic experiential] hallucinations becomes quite clear when the records of the stimulation responses are studied. ... They are reproductions of past experience. ... At operation it is usually quite clear that the evoked experiential response is a random reproduction of whatever composed the stream of consciousness during some interval of the patient's past waking life. (1963, 686-687)

Near the manuscript's end they conclude: "There is within the adult human brain a remarkable record of the stream of each individual's awareness or consciousness. Stimulation of certain areas of cortex, lying in the temporal lobe between the auditory sensory and the visual sensory areas, causes previous experience to return to the mind of the conscious patient" (1963, 692).

That documented cases of phenomenology with sensory qualia and subjective feel induced by cortical stimulation exist at all is interesting and relevant for questions about the neural basis of phenomenal consciousness. But philosophical caution is appropriate here. Penfield and his associates were able to elicit experiential responses in only a minority of temporal lobe epileptics. And these limited results were in epileptic brains, near sites of seizure origin, where electrical activation was admittedly "facilitated" by the organic damage. The evoked experiences were limited not only to (long-term) memory items, but also only to certain types of memories. In the paper's final section, Penfield and Perot (1963) run through a list of memory experiences that were never invoked. Finally, their evidence does not even support the localization of these memory experiences to the site of stimulation. As Penfield and Perot note explicitly, during subsequent interviews days or weeks after their surgeries, patients could recall the experiential responses evoked, even when the site of stimulation had been ablated at a later stage of the surgical procedure! In terms of the neural "location" of subjectively accessible "engrams," the most that these results show is that regions in the temporal cortices, especially ones superior to the first temporal convolution, "play in adult life some role in the subconscious recall of past experience, making it available for present interpretation" by "activating connections with that part of the record of the stream of consciousness in which hearing and seeing are the prominent components" (Penfield and Perot 1963, 689). But Penfield's results were only the beginning. Recent results from three primate research labs carry us further toward induced phenomenology and lessons about the neural basis of sensory qualia and subjective experience.

5 INDUCING PHENOMENOLOGY FROM VISUAL MOTION TO SOMATOSENSORY FLUTTER... AND BEYOND?

5.1 Results from William Newsome's lab

Area MT (*M*iddle *T*emporal cortex) in primates (including humans) is the gateway to the "dorsal" visual processing stream (Figures 4.2, 4.3 above). As discussed earlier in this chapter, this stream processes information about objects' locations and motion leading to actions guided by vision. Both lesion studies and electrophysiological recordings in nonhuman primates have revealed MT's specific role in visual judgments about motion direction. Most MT neurons are *direction selective*, spiking at highest frequency to a visual stimulus moving in a single direction in their receptive fields, a bit less frequently to related directions, and not at all (above baseline spiking rates) to motions unrelated to their preferred direction.[11] Like many cortical regions, MT has a columnar organization. Neurons in a given vertical MT column share similar receptive fields and motion selectivity. These features vary in neurons from column to column, and MT in its entirety realizes a "map" that represents all motion directions at all regions of the visual field (Albright *et al.* 1984).

William Newsome and his collaborators developed a measure that quantifies the strength of a motion stimulus (Salzman *et al.* 1992) (Figure 4.5 below). A pattern of dots appears on a computer screen. The strength of a motion stimulus, expressed as a "percentage correlation," reflects the percentage of dots that are re-plotted on subsequent screens at a fixed spatial interval and direction from their original position. All other dots are re-plotted at the same spatial and temporal intervals but in random directions from their original positions. This re-plotting and the temporal interval between the screens gives the illusion of a motion stimulus, with some percentage of the dots appearing to move in one direction and the rest appearing to move in random directions. For example, in a "50% correlation vertical stimulus," half of the dots on the original screen are re-plotted on later screens at a fixed upward interval, providing the illusion of vertical motion, while the other half are re-plotted randomly.

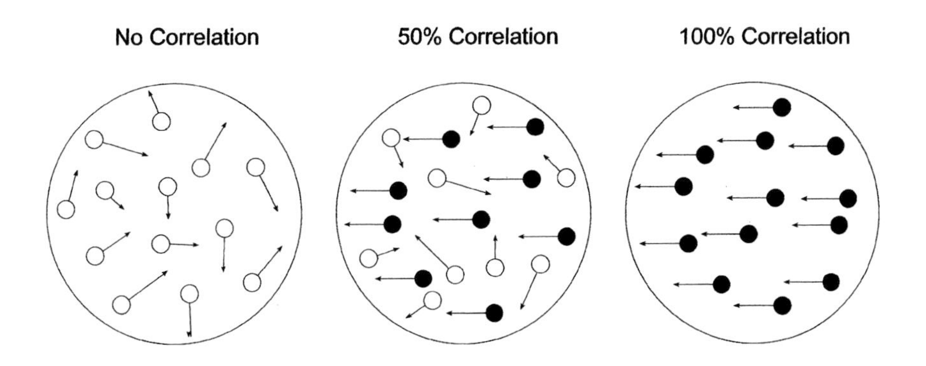

Figure 4.5. Newsome and his colleagues' measure of motion stimulus strength in terms of percent correlation of dots appearing to move in a particular direction (the rest appearing to move in random directions). Reprinted with permission from Salzman *et al.*, figure 1, 2333, copyright 1992 by the Society for Neuroscience.

Newsome's lab also developed a behavioral paradigm in which rhesus monkeys express judgments about motion direction. Their full litany of controls is elaborate but the basic idea is straightforward. The monkey fixates on a central point on a computer screen display and maintains fixation while a visual motion stimulus of a particular strength is presented (a particular percentage correlation in some direction). Both the fixation point and the motion stimulus are extinguished and target lights (LEDs) appear at the screen's peripheries. The "preferred" (Pref) LED is located in the direction (from the original fixation point) of the motion stimulus; the "null" LED is located in the opposite periphery. The monkey indicates its judgment of stimulus motion direction by saccading (moving his eyes quickly) to one of the LEDs. His saccade is his report of apparent (perceived) motion direction. The monkey is only rewarded when he saccades correctly: to the Pref LED in the direction of the percentage correlation motion stimulus. Using standard single-cell electrophysiological recording procedures, Newsome's group first locates an MT neuron's receptive field and preferred motion selectivity. A percentage correlation motion stimulus is then presented only to that neuron's receptive field (as the monkey maintains fixation on the central point). They can then compare the monkey's report about stimulus motion direction across differing strengths (percentage correlation) when electrical stimulation is applied to that neuron through a stimulating electrode and when it is not.

Penfield and his associates induced electrical stimulation through a monopolar silver ball electrode with an area of cortical contact approximately 1.5 square mm. Their typical stimulus was a square wave pulse, 2-5 milliseconds in duration, at 1 to 5 volts and a frequency of 40-100 Hz (cycles per second). The resistance was 10,000-20,000 ohms, yielding a stimulus current that varied between 50-500 milliamperes (Penfield and Perot 1963, 602). Thus vast numbers of neurons were stimulated on a single trial. Newsome's lab microstimulates MT neurons using tungsten microelectrodes with an exposed tip length of 20-30 *microns*. Stimulating pulses are biphasic, each with an 0.2 millisecond duration, with frequency of either 200 Hz or 500 Hz, producing a current of 10 *micro*amperes in amplitude. In one of their publications, Newsome's group reports data from primate motor cortex that a single cathodal 10μA current pulse directly activates neurons within 85 microns of the electrode tip. The number of neurons directly stimulated by the electric current is thus many orders of magnitude smaller than the number directly activated by Penfield's electrodes and pulses.

The percentage correlation measure of stimulus motion strength and their behavioral paradigm permit Newsome's group to plot the proportion of monkeys' reports of apparent motion in stimulated MT neurons' preferred direction as a function of stimulus motion strength. (With MT microstimulation, the "Pref" response is defined as the monkey's judgment that the external stimulus is moving in the MT neuron cluster's preferred direction.) Figure 4.6 represents a monkey's performance with a choice bias slightly in this neuron cluster's preferred direction of motion stimuli. Dots and the sigmoid regression line drawn through them represent the monkey's performance in the absence of electrical microstimulation. When even a small percentage of the dots appear to be moving in this cluster's preferred direction (e.g., > 20% correlation), the monkey correctly judges motion in the preferred direction on nearly every trial (1.0 Proportion Preferred Direction (PD) judgment). When a moderate percentage of the dots appear to be moving opposite this cluster's preferred direction (e.g., < -50% correlation), the monkey correctly judges motion in the null direction on nearly every trial (0.0 Proportion PD). If microstimulation to this direction-selective MT neuron cluster adds signal to the neural processes underlying visual judgments of motion direction, then it will bias the monkeys' reports toward the stimulated neurons' preferred direction. When graphed, this would produce a leftward shift of the psychometric function (Figure 4.6, line A). The monkey then will be more prone to judge motion in the Pref direction, even when fewer of the dots actually appear to be moving that way. If microstimulation adds noise to the neural processes underlying motion judgment, this will exacerbate the monkey's choice bias. When graphed, this would produce nearly constant judgments around the y-intercept of the original function at 0% correlation, with only a slight increase for highly correlated preferred stimulus direction and a slight decrease for highly correlated null stimulus direction (Figure 4.6, line B).

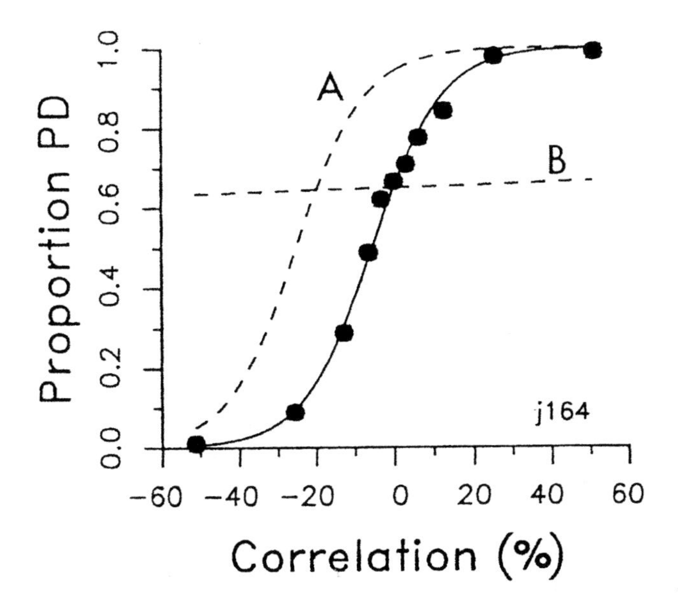

Figure 4.6. Psychometric function relating the proportion of monkey's responses to the MT neuron cluster's preferred motion direction to the percent correlation of the visual motion stimulus. A. If microstimulation to the cluster adds signal to neuronal activity, psychometric curve will be displaced to the left, indicating a higher proportion of responses in the neurons' preferred motion direction to the same strength of visual motion stimulus. B. If microstimulation adds noise, the curve will be replaced by a line nearly perpendicular to the x-axis near the y-intercept value at 0% correlation, with a slightly higher percentage of "preferred" responses to exceptionally strong visual stimuli in the preferred direction and a slightly lower percentage of "preferred" responses to exceptionally strong visual stimuli in the "null" direction. See text for full explanation. Reprinted with permission from Salzman *et al.*, figure 3, 2335, copyright 1992 by the Society for Neuroscience.

Newsome and his colleagues continually observed the "adds signal" results of microstimulation to direction-selective MT neuron clusters, under a variety of percentage correlations (stimulus strengths) and microstimulation frequencies (Salzman *et al.* 1992, especially Figures 4 and 8; Murasugi *et al.* 1993, especially Figures 2 and 5). At nearly every percentage correlation, microstimulation of a direction-selective MT neuron cluster biased the monkeys' saccades significantly to the Pref LED. This bias occurred even in the presence of strong motion stimuli *in the other (null) direction* (e.g., > -50% correlation). Recall that monkeys are only rewarded when they report stimulus motion direction (percentage correlation) correctly. They never receive a reward for their continually incorrect choices under conditions of actual motion stimuli in the null direction and applied microstimulation. Increasing microstimulation frequency (up to 500 Hz) increases the proportion of motion reports in the neurons' preferred direction.

These results lead naturally to the question of what the monkeys are *seeing—experiencing*—during microstimulation trials? Are monkeys *conscious* of motion in the microstimulated neurons' preferred direction, even when the actual motion stimulus is strongly in the opposite direction? Newsome and his colleagues admit that their experiments with monkeys cannot answer this question conclusively. But they don't shrink from offering suggestions, writing: "A plausible hypothesis is that microstimulation evokes a subjective sensation of motion like that experienced during the motion aftereffect, or waterfall illusion. ... Motion therefore appears to be a quality that can be computed independently within the brain and "assigned" to patterned objects in the environment" (Salzman *et al.* 1992, 2352). They are suggesting that *visual motion qualia* are generated in the brain and attached to internal visual representations of external objects. Happily, in ordinary circumstances, our "internal assignments" of features to representations tend to correlate with features and relations of the objects represented. Natural selection was crueler to creatures whose "internal assignments" were more haphazard or skewed. But under appropriate conditions, our internally generated and assigned qualia and the external environmental features can be dissociated. That's what happens, apparently, in Newsome's MT mincrostimulation-motion studies.

Rodolfo Llinás and Patricia Churchland call the general idea behind this suggestion *endogenesis*. They explain: "sensory experience is not created by incoming signals from the world but by intrinsic, continuing processes of the brain" (Llinás and Churchland 1996, x). Incoming signals from sensory receptors keyed to external physical parameters serve only to "trellis, shape, and otherwise sculpt the intrinsic activity to yield a survival-facilitating, me-in-the-world representational scheme" (ibid.). Natural selection—and hence adequacy for exploiting an available environmental niche, not truth—determines a scheme's success.

Microstimulation motion effects are not specific to nonhuman primates. Newsome and his colleagues remark that "it has recently been reported that crude motion percepts can be elicited with electrical stimulation of the human parietal-occipital cortex" (Salzmann *et al.* 1992, 2354). Their results are continuous with Penfield's pioneering studies (reported in the previous section). Nor are microstimulation effects specific to visual motion. Newsome and his collaborators have also succeeded in affecting judgments of stereoscopic depth. One important depth cue for creatures with overlapping visual fields (like primates) is *binocular disparity*, small differences in position between images of an object formed on the two retinas. Neurons responding selectivity to binocular disparity have been found throughout primate visual cortex, including in area MT. There they are found in clusters, typically 200-300μm in diameter, surrounded by neurons with little or no disparity selectivity. Each neuron in a cluster responds optimally to a common range of preferred binocular disparities. By finding such clusters using standard microelectrode search techniques and then inducing electric microstimulation into their centers, Newsome and his collaborators have been able to affect judgments of binocular disparity similar to those they elicited on judgments of motion direction (DeAngelis *et al.* 1998).

Monkeys begin the depth-discrimination task by fixing their gaze on a point in the center of a computer screen display. A random-dot pattern is presented within a circular aperture roughly the size and location of the multiunit receptive field of the MT neurons to be microstimulated. The dots, arranged as red/green anaglyphs viewed at the appropriate distance through red/green optical filters, give the illusion of depth at particular degrees of horizontal binocular disparity.[12] Similar to their measure of motion stimulus strength, Newsome and his collaborators measure depth stimulus strength in terms of percentage of binocular disparity correlation. Some percentage of the dots (the "signal") shown to the right eye (through the colored filter) are displayed with a dot shown to the left eye to produce identical degrees of horizontal disparity and the illusion of constant depth (either closer or further away than the fixation point); the rest of the dots ("noise") in the receptive field aperture are presented to produce random disparity. Monkeys report judgments of near or far depth by saccading to an appropriate LED. A trial begins when the fixation point appears and the monkey fixes his gaze. Dots then appear in the receptive field aperture to reflect some strength of depth stimulus. On half the trials, selected at random, microstimulation to the disparity-selective MT neuron cluster begins when the visual stimulus appears. After one second the visual stimulus is extinguished (along with microstimulation, if it occurs on the given trial) and the LED target lights appear. The monkey saccades quickly to the appropriate target to reflect his judgment of the stimulus's near or far disparity. Rewards are given on all and only the trials when the monkeys' response correctly indicates the depth stimulus ("signal").

Just as they found with MT microstimulation and motion stimuli, Newsome and his collaborators found that microstimulation adds signal to monkeys' judgments about stereoscopic depth. In one MT cluster tuned to "near" disparity displays, at ≥ 50% binocular correlation—where one-half or more of the "signal" dots are at the neurons' preferred degree of horizontal disparity (i.e., the degree that generates maximal spiking frequency in these neurons under normal viewing conditions)—and no microstimulation, monkeys gave a correct judgment (saccaded toward the "Near target" LED) on roughly 80% of the trials. Similarly, at ≤ –15% correlation, when 15 percent or more of the signal dots were opposite these neurons' preferred disparity, and no microstimulation, monkeys correctly saccaded away from the "Near target" LED on every trial. However, with microstimulation to this "near-preferring" cluster, monkeys gave "Pref" judgments on 80% of the trials with -30% binocular correlation—when thirty percent of the dots displayed disparity *opposite to* the microstimulated neurons' preferred disparity and the rest displayed random disparity. "Preferred" judgments reached 100% with microstimulation on trials with 0% horizontal disparity (i.e., when every dot displayed random disparity). This was a robust and not atypical result (DeAngelis *et al.* 1998, Figure 3b, d, p. 679). Newsome and his collaborators report 43 statistically significant differences on microstimulation versus no microstimulation trials in 65 depth disparity experiments on two monkeys. On 42 of these 43 cases, microstimulation biased monkeys' depth disparity judgment toward the stimulated cluster's preferred depth disparity (DeAngelis *et al.* 1998). Statistically significant results were limited to clusters that displayed moderate to strong disparity selectivity. As before, the intriguing suggestion is that activity in MT depth-selective neurons generates visual depth qualia that get "assigned" to representations of external visual stimuli. Subjective phenomenal conscious experience with depth qualia seems to be induced by highly specific electric microstimulation to tiny clusters of depth-selective neurons in area MT.

5.2 Results from Kenneth Britten's lab

Nor are microstimulation effects specific to only one lab. Changing patterns of visual motion on the retinas as we move through space, called "optic flow," provides a rich source of information about direction of self-movement or "heading." Earlier work with rhesus monkeys indicated that the medial superior temporal area (MST) (Figures 4.2 and 4.3 above) contains neurons selective for optic flow information and for stimuli that simulate the visual effects of self-motion. Neurons tuned to leftward or rightward heading directions are arranged in clusters spanning regions up to 500μm. Britten and van Wezel (1998) presented rhesus monkeys with visual displays that simulated a cloud of dots at a visual depth from 1-10m. All dots were re-plotted at a particular angle and distance to provide the illusion of self-motion through space at a particular angle and direction ("heading") (Figure 4.7A, B). A trial begins when a red fixation point in the center of the visual display is illuminated and the monkey fixes its gaze upon it. Throughout the trial's duration, the fixation point either remains at that location or moves at a constant velocity to the left or right. The dot field appears shortly after the fixation point. Dot re-plotting begins immediately to simulate optic flow in a specific leftward or rightward horizontal heading angle. 0° corresponds to "directly ahead," negative degrees correspond to leftward heading, and positive degrees correspond to rightward heading. On half of the trials microstimulation at 20μA amplitude, 200 Hz frequency, begins at the same time that the dot field appears. Stimulating electrodes have been inserted into the center of MST clusters of at least 250μm in which all neurons are tuned to a similar leftward or rightward heading direction and angle. One second later the dot field is extinguished, microstimulation ceases (if it occurs on the trial), and target lights are illuminated, one in the angle and direction of the heading stimulus, the other at the same angle but opposite heading direction. Monkeys are rewarded if they saccade to the target in the direction of the heading stimulus. The moving fixation points on some trials forces the monkey to make smooth pursuit eye movements during the stimulus period. MST neurons are known to display activity correlated with smooth pursuit eye movements and appear to compensate their heading tuning for distortions in optic flow produced by smooth pursuit.

Once again, results indicate that microstimulation adds signal to the neural processes underlying judgments of heading direction. In one experiment, the proportion of rightward heading choices made by the monkey was plotted against degree of horizontal heading presented in the dot field. The neurons in the region of the stimulating electrode tip were selective for rightward optic flow and leftward heading. This monkey exhibited a baseline bias, perceiving "directly ahead" at roughly -5° (to the left) of the display's geometric center. When this cluster of leftward heading-tuned MST neurons was microstimulated and the visual display indicated leftward heading at -4° (which to this monkey correlated to $+1^o$ rightward heading due to his baseline bias), the monkey judged that the heading was rightward only slightly more than 1 in 4 trials; without microstimulation to the leftward heading-tuned neuron cluster under these same viewing conditions, the monkey judged rightward heading in 8 of 10 trials. At -1° (leftward) heading stimulus (which to this monkey correlated to $+4^o$ rightward heading), microstimulation to the

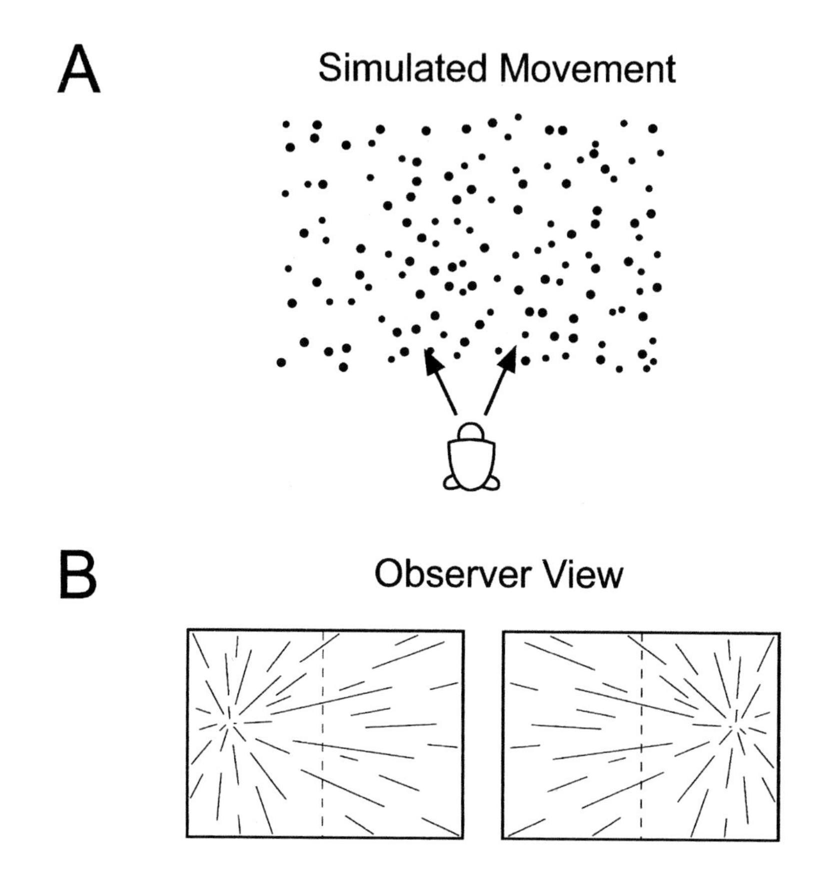

Figure 4.7. Heading direction visual stimulus in Britten and Wezel's MST microstimulation studies. A. The simulated "virtual" depth stimulus and heading direction of the monkey with respect to it, as seen from above. B. Appearance of left- and right-heading stimulus, as seen from the observer's perspective. Reprinted from *Nature Neuroscience* **1**, K. Britten and R. van Wezel, copyright 1998, pages 59-63, with permission from Nature Publishing Group.

leftward heading-tuned neuron cluster dropped the monkey's judgment of rightward heading direction from 100% (without microstimulation) to just below 60%, only slightly above chance. Results were even more dramatic when the fixation point moved leftward during the dot field display, forcing the monkey into leftward smooth pursuit eye movements. Even when the dot field indicated a rightward heading of +4°, when without microstimulation the monkey indicated rightward heading on 75% of the trials, microstimulation of this leftward heading-tuned MST cluster biased the monkey to choose the *leftward* heading target dot in 90% of the trials (Britten and van Wesel 1998, Figure 2, p. 60). As in Newsome's studies, monkeys are never rewarded for incorrect choices of actual heading stimulus. Here we have the visual phenomenology of self-motion through space seemingly induced by highly specific electrical microstimulation. Heading direction in a cloud of moving dots giving the illusion of 10m in depth is complex visual phenomenology indeed to induce by stimulating a cluster of neurons between 250-500μm, especially under conditions of *opposite* visual presentation.[13]

5.3 Results from Ranulfo Romo's lab

Nor are microstimulation effects limited to *visual* stimuli. Ranulfo Romo and his collaborators trained rhesus monkeys to distinguish differences in frequency between two flutter stimuli delivered to a fingertip site. Humans report sensations of "flutter" when mechanical vibrations between 5-50 Hz are applied to the skin. Such stimuli activate neurons in primary somatosensory cortex (area 3b of S1, see Figure 4.2 above) whose receptive fields include the stimulation site. "Quickly Adapting" (QA) neurons are strongly activated by periodic flutter vibrations and fire with a probability that oscillates exactly at the input frequency. In other words, their mean firing rate correlates directly with the frequency of the mechanical vibration applied to their receptive fields (appropriate portions of the skin) (Mountcastle *et al.* 1990). These neurons are also arranged in columnar clusters that share similar receptive fields. All these properties make them a convenient target for microstimulation studies.

In their first study (Romo *et al.* 1998), a trial began when a mechanical probe was lowered onto a monkey's restrained hand, indenting slightly the glaborous skin of one fingertip (in lay terms, the fingerpad). The monkey then places its unrestrained right hand onto an immoveable key within 1 second. After a brief delay period (1.5-3 seconds) the mechanical probe oscillates at the "base" frequency for 500 milliseconds. This is followed by a brief delay (1-3 seconds), after which either a second mechanical stimulus (the "comparison") at either a higher or lower oscillatory frequency than the base stimulus is delivered to the restrained hand or microstimulation at a frequency that corresponded to a higher or lower mechanical stimulus is delivered to the QA neuron cluster in S1 whose multi-unit receptive field includes the stimulus site. The monkey indicates detection of the end of the "comparison" frequency (mechanical or cortical microstimulation) by releasing the key within 600 milliseconds, and whether the "comparison" frequency was higher or lower than the "base" by pushing one of two buttons with its free hand. Monkeys are rewarded on all and only trials in which their comparative judgment about whether the "comparison" stimulation was of higher or lower frequency than the "base" is correct. By comparing performances on mechanical versus microstimulation "comparison" stimuli, Romo and his collaborators sought to discover "whether the animals could interpret the artificial signals [microstimulation "comparison" frequencies] as flutter" (Romo *et al.* 1998, 388).

As long as "base" and "comparison" stimuli consist of two current pulses with amplitude > 65μA, monkeys achieve over 75% correct for both mechanical and microstimulation "comparison" frequencies. This result obtains even when the "comparison" frequency differs from the "base" by only 8 Hz. They show no statistically significant differences at these frequencies whether the comparison frequency is actual mechanical stimu-lation to the fingertip or cortical electrical microstimulation. When the base frequency is held constant over trials at 20 Hz, monkeys make correct judgments about comparison frequencies better than 75% of the time when these frequencies are ≤ 15 Hz or ≥ 25 Hz, with no statistically significant difference between mechanical stimuli and microstimulation (Romo *et al.* 1998, Figure 2, p. 388). Romo and his collaborators conclude: "The monkeys were consistently able to extract the comparison frequency from the artificially induced sensation" (Romo *et al.* 1998, 388).

What about our issue of "induced phenomenology"? Romo *et al.* are not timid here, writing:

> Animals continuously switched between purely mechanical and microstimulation conditions with almost identical performance levels. Such high accuracy, based on the interaction between natural and artificially evoked activity, is consistent with the induction of a sensory percept. ... Thus the microstimulation patterns used may elicit flutter sensations referred to the fingertips that are not unlike those felt with mechanical vibrations. (Romo *et al.* 1998, 399-390)

They also report that on seven experimental runs during this study, they induced a combined mechanical and microstimulation comparison stimulus. The mechanical component had a lower frequency than the base stimulation while the microstimulation was higher. Despite the actual stimulus induced at the fingertip and no reward given for "incorrect" mechanical judgments, monkeys judged the (combined) comparison stimulus as higher frequency in 348 of 400 such trials.

In a subsequent study, Romo and his collaborators reversed the base and comparison stimuli from their original study (Romo, *et al.* 2000). All other parameters remained as described above, but now on half the trials the *base stimulus* was mechanical vibration at the fingertip while on the other half it was instead cortical microstimulation to area 3b in S1 (again to QA neuron clusters whose receptive fields contained the fingertip stimulation site). The comparison stimulus was mechanical stimulation at either a higher or lower frequency. In their previous study, monkeys had to be capable of comparing the result of microstimulation with the base frequency, with the latter represented and stored in working memory through normal means (beginning with actual mechanical stimulation of the fingertip). In this second study, however, if the monkeys are to succeed with microstimulation "base" stimuli comparable to their performance with actual mechanical "base" stimuli, then cortical microstimulation alone must engage the entire range of cognitive processes involved, from sensation through working memory and comparative decision-making. So besides its intrinsic scientific interest, this study is doubly relevant for our question of induced conscious experience because of the tight connection many recognize (and I stressed in sections 1 and 2 of this chapter) between working memory and consciousness.

Even when base and comparison frequencies differed by as little as 4 Hz, monkeys were able to respond correctly on 75% of the trials about which frequency was lower. There were no statistical differences between performances on trials with mechanical stimulus or microstimulation base. As a control, Romo and his collaborators tested monkeys when both base and comparison frequencies were microstimulations alone, comparing results to cases of similar base and comparison frequencies where both were actual mechanical stimuli. In the former cases, there were no actual mechanical stimulations to induce sensory, working memory, or comparative decision-making processes in the normal fashion. Yet monkeys performed nearly identically in the two types of cases (although there was more variance within sessions with the purely artificial base and comparison stimuli).

Is there a foreseeable limit on this latest wave of cortical microstimulation studies? Newsome and his collaborators have long sought similar results with color stimuli. In their first paper reporting microstimulation results with visual motion stimuli, they write:

> A natural extension of this work is to apply the same basic approach to the study of circuits that mediate aspects of visual perception other than motion. In principle, the microstimulation technique is applicable to the analysis of function in any circuit in which neurons with similar physiological properties are segregated into columns or large clusters. ... Given present physiological knowledge, appropriate candidates for future investigation are circuits that encode orientation, *color*, and disparity. (Salaman *et al.* 1992, 2353; my emphasis)

However, color continues to pose formidable technical difficulties (Newsome, personal correspondence). Going beyond sensory stimuli, Liu and Newsome (2000) recently have raised the possibility of microstimulating the appropriate neurons involved in the working memory and comparative decision-making aspects of tasks like Romo's. We saw in sections 1 and 2 above that cells with "working memory fields," that fire during short delay periods (up to 12 second), have been found in prefrontal cortex. Might microstrimulation to clusters of cells sharing working memory fields and properties induce causally efficacious "working memories"? As Liu and Newsome put this question, "might it be possible to influence or change the monkey's memory by electrically stimulating such neurons?" (2000, R600). Current physiological knowledge has not yet established that these neurons are grouped anatomically into columns or clusters with others sharing similar activation properties. However, as Liu and Newsome note, "only a few years ago the complexity of the cerebral cortex would have led most sensory physiologists to declare Romo and colleagues' current microstimulation experiments a fantasy" (2000, R660). They insist that "for now, all bets are off until the experiments are actually tried" (ibid.).

There is an additional reason for why these microstimulation results from Penfield's through Romo's are central to a ruthlessly reductive cellular neuroscience of conscious experience. I stressed throughout Chapters Two and Three that the capacity of experimenters to intervene directly on lower level mechanisms to generate specific, measurable behavioral effects is crucial for claiming an explanation, and hence a reduction. Microstimulation procedures accomplish exactly this result for "induced phenomenology." Experimenters directly activate tiny clusters of stimulus-specific sensory neurons and generate observable behavior assumed to be guided normally by visual or somatosensory experiences. These direct cellular manipulations that induce sensory experience are consistent with the general explanatory demand recognized in current mainstream neuroscience. Don't simply find neuronal or intra-neuronal activity correlated with a cognitive task; *invoke* the specific behavior by directly manipulating the hypothesized cellular or molecular mechanisms.[14]

6 THE STRANGE CASE OF PHENOMENAL EXTERNALISM[15]

It has been more than twenty-five years since Hilary Putnam first took philosophers to "Twin Earth." Twin Earth is just like Earth, down to molecule-for-molecule duplicate beings, except for one feature in the linguistic or physical environment. In Putnam's (1975b) original fable, the single difference was water's molecular structure. On Earth, water is H_2O; on Twin Earth, the clear, tasteless liquid in rivers, streams, swimming pools, rain showers, drinking fountains, etc., is XYZ. Putnam sought to show that "meanings ain't in the head." Since meaning determines reference, your utterances of "water" refer to H_2O, and Twin-your utterances refer to XYZ, the homonyms must have different meanings. But nothing is different inside your and Twin-your heads, so whatever individuates meanings ain't there.[16]

Twin Earth's legacy in philosophy is equally legendary. Tyler Burge (1979) developed footnote 2 of Putnam (1975b) into a full-blown theory of mental content, and philosophical orthodoxy decided that whatever individuates mental content "ain't in the head," either. *Content externalism*, advocating *wide content*, became the rage. Features external to cognizers "individuate" their mental contents. Content "extends out into the world." In the wake of the consciousness craze that swept philosophy throughout the 1990s, it was only a matter of time until prominent philosophers of mind extended content externalism and the Twin Earth fantasy to phenomenal or qualitative content.

Phenomenal externalism holds that the environment external to an individual's receptor surface "individuates" the qualitative content (qualia) of his or her sensory experiences. Features of the external environment, literally, distinguish qualitative states from one another (Dretske 1996; Lycan 1996). A common intuition pump for this view is ... Inverted Earth! First suggested by Gilbert Harman (1982), the Inverted Earth fantasy reached fruition in Ned Block's ([1990] 1997) essay.[17] Block writes:

> Inverted Earth differs from Earth in two respects. First, everything has the complementary color of the color on Earth. The sky is yellow, grass is red, fire hydrants are green, and so on. I mean everything *really* has these oddball colors. ... Secondly, the vocabulary of the residents on Inverted Earth is also inverted: If you ask what color the (yellow) sky is, they truthfully say "Blue!" ... If you brought a speaker of the Inverted Earth dialect to a neutral place (with unknown sky color, unfamiliar vegetation, and the like) and employed a team of linguists using any reasonable methods to plumb his language, you would have to come to the conclusion that he uses 'red' to mean what we mean by 'green,' 'blue' to mean what we mean by 'yellow,' and so on. ([1990] 1997, 682)

Of course, no saga worthy of Putnam's legacy could stop here. The next step is ... Secret Transplanetary Relocation! Block continues:

> A team of mad scientists knocks you out. While you are out cold, they insert color-inverting lenses in your eyes, and change your body pigments so you don't have a nasty shock when you wake up and look at your feet. They transport you to Inverted Earth, where you are substituted for a counterpart who has occupied a niche on Inverted Earth that corresponds exactly (except for colors of things) with your niche at home. You wake up, and since the inverting lenses cancel out the inverted colors, you notice no difference at all. ([1990] 1997, 683)

"What it is like" for you on Inverted Earth, with the lenses and body pigment alterations, seemingly doesn't change from "what it is like" for you on Earth. When you look up at the yellow sky, you have a brilliant "Carolina blue" quale (as they say in Lycan's and Dretske's current abode). This experience is just like the quale you would have minus the lenses if you looked up at the blue sky from Earth.

However, the phenomenal externalist must balk at the intuition just expressed. He or she must insist that the qualitative contents of your visual experiences on Earth and Inverted Earth (with the implanted lenses) must *differ*. Dretske (1996) is explicit on this point. He claims that nothing prevents one who uses Twin Earth intuitions to defend externalism about linguistic meaning and mental content from using them to defend phenomenal externalism. "Just as we distinguish and identify beliefs by what they are beliefs about, and what they are beliefs about in terms of what they stand in the appropriate relation to, so we must distinguish and identify experiences in terms of what they are experiences of" (1996, 145).[18] The radical nature of Dretske's proposal is apparent in his slogan: "The experiences themselves are in the head ... but nothing in the head ... need have the qualities that distinguish these experiences" (1996, 144-145). Earth-you and altered-and-transported-to-Inverted-Earth-you have conscious visual states *with different phenomenal content—different qualia*—as you (plural) gaze up into your (plural) respective skies. Earth-yours is blue; it is an experience *of* a blue sky. Altered-and-transported-to-Inverted-Earth-yours is yellow; it is an experience *of* a yellow sky.

Like many philosophical fantasies that began with good intentions, Inverted Earth has spawned a host of confusing philosophical exotica. Block ([1990] 1997) and Lycan (1996) invented "Strange Qualia" and "New Strange Qualia," to name just two, and I'm sure that more are on offer in this growing philosophical literature. In their review of Lycan (1996), Tom Polger and Owen Flanagan (2001) liken the qualia literature spawned by Block's and Lycan's exchanges over Inverted Earth to a massive crash during an car race that continues on at full speed. They warn: "For those who can identify working bits among the mangled wreckage, there are some excellent parts for the taking. But try not to get hurt" (2001, 120-121). Lycan doesn't balk at this assessment, writing in his reply that "Polger and Flanagan's vivid description of the Block-Lycan imbroglio over Strange Qualia and New Strange Qualia recalls exactly what it felt like to be embroiled in it, especially the part about trying to describe the accident as we saw it on all the monitors and, of course, fixing blame" (2001, 130-131, footnote 1).[19] At the risk of oversimplifying a complicated philosophical matter, my diagnosis for why this discussion imploded has to do with the fantastical nature of the Inverted Earth fantasy. Garbage in, garbage out, as the saying goes, and when the input is as unconstrained by reality as Body Inversion and Secret Transport to Inverted Earth scenarios, small wonder that confusing philosophical exotica and esoteric discussions soon arose.

Happily, the microstimulation results surveyed in section 5 of this chapter give us real (Earthly) scientific analogs to the Inverted Earth fantasy. Consider the key features of an Inverted Earth scenario, in which the Earthling individual has been transported with the lenses and body pigment changes:

- Same (internal) *brain state* (because of the inverted lenses);
- Same perceptual judgment ("Still blue");
- Different external stimulus (blue sky versus yellow sky).

The issue between the phenomenal internalist and externalist is whether the contents of the qualitative states—the qualia—count as the same or different across these features. *But these are exactly the features of the normal presentation without microstimulation versus the microstimulation plus opposite stimulus in the studies surveyed above.* Consider the Newsome lab's motion direction study (although any of the others would illustrate this point equally well). The "Earth" analogy is the case where the monkey is presented with a stimulus with a percent correlation of dots moving in a particular direction, say 90° left, and no microstimulation. The "translocated Inverted Earth" analogy is the case where the monkey is presented with the opposite stimulus, e.g., the same percent correlation of dots moving 90° right, plus microstimulation to a left-preferring cluster of MT motion selective neurons. The microstimulation is analogous to the inverting lenses. Across these two experimental cases we have, analogous to the above:

- the "same" brain states, one induced by the external leftward motion stimulus, the other induced by microstimulation;

- the same judgment about motion direction, since the monkey in both cases will saccade consistently to the leftward or "Pref" target light;
- different external environmental stimuli, namely, percent correlation motion to the left in the former case and to the right in the latter.

We thus don't need the fantastic fictional example. Microstimulation studies being pursued in real laboratories right here on Earth capture exactly the features that Inverted Earth with Secret Transplanetary Relocation were designed to illustrate. We can thus ask the phenomenal internalist and externalist to reflect on real science. Behold, a methodological gain in the philosophy of consciousness!

Block ([1990] 1998), incidentally, anticipated the possibility of an empirical case analogous in the important ways to Inverted Earth, although he has nowhere suggested that current microstimulation research provides this. When considering an objection to his Inverted Earth criticism of intentionalist/functionalist accounts of qualia, namely that "the boundary between the inside and the outside should be moved inward, and inputs should be thought of in terms of the color of the light hitting the retina, not in terms of colored objects," Block replies: "I mentioned an alternative to the inverting lenses, one that seems to me more physically plausible than the lenses, namely, a neural inverter behind the retina. ... I don't know any reason why it shouldn't be in principle possible for a miniaturized silicon chip to register these impulses and substitute transformed impulses for them" ([1990] 1998, 688). Substitute "tungsten electrode" for Block's fantasized silicon chip, implant it well back "behind the retina" (into the appropriate cortical region for the stimulus at issue) and you have the *actual* scientific method of current microstimulation studies.

However, more than just a gain in philosophical method seems to be on offer here. The neuroscientists doing the microstimulation studies surveyed above commonly suggest that the monkeys' *subjective conscious experiences were similar* across the microstimulation and no-microstimulation cases, despite the different external stimuli. These scientists don't use philosophical jargon like "qualia," "phenomenal content," and "individuation," but we can easily translate the suggestive comments surveyed in the previous section into these terms. According to these neuroscientists, their results show that what matters for qualitative sensory experience is *what goes on in the brain*, not what goes on in the external environment. Normally, external stimuli correlate systematically with internally generated qualia. But as the microstimulation studies consistently suggest, the two can be dissociated. When they are, monkeys consistently judge stimulus qualities based on the preferred stimuli of the neurons being microstimulated, not the actual features of the external stimuli. The natural interpretation is that the internally generated subjective qualia across microstimulation and no-microstimulation cases *share the same features*. Any generalized skepticism about "actual phenomenology" being induced electrically in higher primates is softened by the variety of stimulus types and modalities for which these microstimulation results have been gathered and, most importantly, from linguistic reports from one type of higher primate—humans—undergoing the Penfield procedure. The science points in favor of internalist intuitions about phenomenal, qualitative content.

I round off this discussion by returning to Lycan's comment about the "car crash" quality of the Block-Lycan debate. After agreeing with Polger and Flanagan's assessment (see footnote 19 of this Chapter), Lycan ends his discussion with the following suggestion:

> I would point out, though, that the *issue* between Block and me was clear: I maintain that mental states and events (as such) have no metaphysically interesting properties save their functional properties and their representational properties. Block disagrees, insisting that phenomenal states have special properties of a third sort. The mayhem that followed was a matter of my trying to understand exactly what sorts of property those are supposed to be, and (as you might guess) of the two of us simultaneously competing to distribute the burden of proof. (2001, 131, footnote 1)

But in at least one passage Block offers a suggestion about this "third sort" of property, and the scientific lessons I surveyed in the previous two sections help fill it out. He writes: "I take the view that thought is functional, but that the best bet for a theory of qualia lies in neurophysiology" ([1990] 1998, 693, footnote 29). Block's qualia need not be some metaphysical "third property," as Lycan suggests, because even functionalists about intentional states like Lycan and Block presume (at least token) physiological realizers for all types of mental states. But Block is suggesting that the physiological properties of phenomenal states exhaust their qualitative contents. The microstimulation studies suggest such an account. What is the motion quale of my visual experience as the lobbed football moves across my visual field? It is the activity in motion-selective MT neurons, especially those spiking most frequently to actual motion in my visual field at that speed, direction, and location. Dissociate the internal and external events—lob the football across my visual field in a different direction while microstimulating a small cluster of MT neurons selective for motion stimuli in the first direction—and I'll have the same motion quale as before. Only now, I'll be a lot less effective as a receiver on your flag-football team. A similar account can now be offered for visual depth qualia, visual heading qualia, and somatosensory flutter qualia, with the promise by the neurophysiologists carrying out these studies of more coming in the future.

And yet we'll still hear the clamor: "But action potential frequency patterns in cortical neurons don't *seem* like sensory qualia! How can neuroscience bridge that gap?" Although the available evidence—i.e., the philosophical literature on consciousness—suggests that nothing changes opinions on either side of this clamor, I'll end this chapter with evidence that prompts me to dismiss this clamor. This evidence reflects an attitude toward "the hard problem" from the hard-core Society for Neuroscience crowd.

7 THE "HARD PROBLEM" AND THE SOCIETY FOR NEUROSCIENCE CROWD

There are neuroscientists who think of the brain as "just another organ" or "just another piece of biological tissue." However, many pursue the discipline for reasons that historically have motivated humanists, and are not afraid to express these motives in print. A nice example is this passage from the introductory chapter of Gordon Shepherd's influential textbook: "As we grow older, we experience the full richness of human behavior—the ability to think and feel, to remember and create—and we wonder, if we have any wonder at all, how the brain makes this possible" (1994, 3). This is not the ranting of some left-field crank, but rather from the current editor of the *Journal of Neuroscience*. Similar passages can be cited many-fold. Many bench neuroscientists aren't philosophical philistines.

These passages won't satisfy some philosophers, however, who remain jealous guardians of the "qualitative" and "subjective" aspects of Mind. They assume that only they, and perhaps a handful of theoretically minded psychologists, grapple seriously with "what it is like" to be a conscious, mindful human being. They assume that these features are beyond neuroscientists' professional grasp and serious interest. But they are wrong. Consider the following passage from William Newsome, commenting on microstimulation studies from his lab (surveyed above in section 5.1 of this chapter):

> I believe *the nature of internal experience matters* for our understanding of nervous system function ... Even if I could explain a monkey's behavior on our task in its entirety (in neural terms), *I would not be satisfied* unless I knew whether microstimulation in MT *actually causes the monkey to see motion*. If we close up shop before answering this question and understanding its implications, we have mined silver and

> left the gold lying in the tailings. (Gazzaniga 1997, 65-66; my emphases)

Yet Newsome asks for no special discipline or methodology to address this question. He sees no shortcuts around a broadly empiricist, reductionist path, writing: "For the time being ... I suspect we must feel our way towards these ambitious goals from the bottom up, letting the new light obtained at each level of inquiry hint at the questions to be asked at the next level" (Gazzaniga 1997, 67).[20]

Zealous guardians of "the hard problem" in the philosophy of consciousness should loosen up. Philosophers aren't the only ones respectful or in pursuit of the full glory of Mind. If we are to trust neuroscientists pursuing "mainstream" research on cellular and molecular mechanisms, this problem is not beyond their professional interest or training. And any philosophical position that insists on *a priori* grounds that "it must be, based on the very nature of reductionistic science," would profit from taking a look at actual, mainstream, professional-scientific-journal-publication research. The lesson I've urged throughout the last three chapters is that the philosophical lessons lie there, even for "the hard problem" of consciousness. And to my lights, pursuing those lessons there sure beats chiming in with one more intuition about worn philosophers' fantasies like Twin Earth, inverted spectra, dancing qualia, and Mary the colorless visual neuroscientist.

NOTES

[1] One set of results I will not discuss is object-specific activity in neurons comprising regions of the "ventral" visual stream (see Figures 4.2 and 4.3 below). My ignoring these experiments and results is not due to their lack of scientific interest or philosophical import, but rather because they have already been presented to philosophers of consciousness in an admirable paper by Jesse Prinz (2000).

[2] For example, Dennett (1991) gives an extended argument against theater metaphors for the conscious mind.

[3] With one exception: Funahashi *et al.* (1989) report some differences in PS neuron response properties compared with frontal eye field (FEF) neurons during the ODR task. Since the output of a preponderant number of FEF neurons is known to code for oculomotor dimensions of an upcoming saccade, these differences are suggestive against the alternative explanation. I discussed FEF pre-saccadic activity briefly, with references, Chapter Three, section 3.2 above.

[4] On the Stroop Task, subjects are presented with a list of color words ('blue,' 'green,' 'red,' etc.) that appear in color, but never the same color as the color word's reference. For example, the word 'red' appears in blue, 'green' in red, and so on. One task is to name the color in which the word appears as quickly as possible, which requires subjects to inhibit reading the color name. The Wisconsin card sorting task presents subjects with four stimulus cards, each with

designs that differ in color, form, and number of elements. There are numerous variations, but in one the first card might have one pink heart, the second two yellow moons, the third three orange stars, and the fourth four green clovers. The rest of the cards in the deck have alternative combinations of these colors, forms, and numbers. Subjects must sort each remaining card in the deck by placing it in front of one of the four stimulus cards. The only hint subjects are given is an indication of whether a given sort is correct or incorrect. Unannounced to the subject, "correct" and "incorrect" are determined in the following way. Color is the first solution. As soon as the subject indicates that he or she has figured this out, the solution suddenly changes (without warning or explicit announcement) to form. To succeed, subjects must inhibit sorting the cards on the basis of color and switch to form. Once the subject has switched successfully to form, the solution changes (without warning or explicit announcement) to number of elements. Later it will become color again, and so on. Both tasks are used extensively in neuropsychological investigations of frontal lobe damage, as patients with certain types of frontal damage perform very poorly on them. Incidentally, nonpatient controls regularly express subjective difficulty with these tasks and the need for explicit, conscious concentration to perform well. For a "textbook" description of these tasks and their neuropsychological use, see Kolb and Whishaw 1996, chapter 14.

[5] Pyramidal cells are a type of neuron, so named because of the shape of their cell bodies. They are the primary type of "working memory" neurons in prefrontal cortex.

[6] In keeping with the theme in reductionist neuroscientific research emphasized in Chapter Two, section 5.2 above, they note that this result is consistent with some early results using D1 mutant knock-out mice.

[7] For those worried that my talk of the causal effects of explicit conscious attention on single neuron activity borders on dualism or mystery, be comforted. Toward the end of this section I'll discuss a neurobiological explanation of these effects that is under active development and investigation.

[8] This isn't just an arbitrary intuition on Chalmers' part. He has a much-discussed theory of reductive explanation, most fully developed in the early chapters of his (1996) book.

[9] Thanks to John Symons for separating these two projects and emphasizing the importance of the first.

[10] Case and page numbers cited in the next three paragraphs refer to the case histories in Penfield and Perot (1963).

[11] Thus their responses likewise are fit by Gaussian curves (Figure 4.4A above), where values on the x-axis represent motion direction parameters.

[12] Fans of cheesy 1950s 3-D horror and science fiction movies are familiar with anaglyphs. Remember the cardboard glasses with red and green plastic lenses? Anaglyphs are stereoscopic motion or still pictures whose right component, usually red in color, is superimposed on the left component, usually green, to produce a three-dimensional effect when viewed through differently-colored optical filters over the two eyes.

[13] Britten and van Wezel report a fair amount of heterogeneity in their microstimulation effects. In a few cases, microstimulation of MST clusters biased the monkeys' choices *opposite of* the heading-direction tuning of the neurons in the cluster, i.e., stimulating a cluster tuned to left heading direction biased the monkey choice toward the right heading target. They offer explanations for this puzzling effect, one being that the visual display far exceeds the receptive field of the stimulated neurons. The monkeys probably use cues from the entire stimulus, not

just from within the receptive fields of the stimulated neurons. Lateral interactions within MST could then on occasions override the microstimulation effects. They also note that heading computations might be exceedingly complex, keyed to more dimensions of the visual stimulus than the ones controlled for in this study. Given the obvious complexity of this type of visual stimulus and the neuronal computations processing it (even within a single region like MST), that Britten and van Wezel (1998) achieved results as robust as they did is fascinating from our perspective of phenomenology induced by neuronal microstimulation.

[14] Thanks to Huib Looren de Jong and Maurice Schouten for continually impressing this worry about explanation concerning the scientific evidence presented in Chapters Two through Four.

[15] Thanks to Tom Polger for discussions that helped me clarify the views and arguments of this section.

[16] Twin-you is your molecule-for-molecule *doppelgänger*—okay, ignore the fact that so much of you consists of *water*!

[17]However, Block adapted Inverted Earth from Harman to *criticize* "functionalist/ intentionalist" theories of qualitative experience. His targets were early prototypes of representationalist theories prominent these days, and closely allied with phenomenal externalism.

[18] The necessity modality in the last clause of this quotation must be viewed in larger context. Throughout his (1996), Dretske is careful to point out that he is urging the *availability*, not the truth, of phenomenal externalism. Lycan (1996) is a bit bolder, albeit less pithy.

[19] Lycan doesn't end the footnote here, however. Despite the admitted car crash quality of the Block-Lycan exchange, he still thinks that he can put his finger on the issue that really separates the two views of qualia. I'll say more about his additional remarks at the end of this section.

[20] Notice that Newsome's reductionism doesn't collapse levels, at least not methodologically, as many anti-reductionist philosophers fear.

REFERENCES

Abel, T., P. Nguyen, M. Barad, T. Deuel, E.R. Kandel, and R. Bourtchouladze (1997). "Genetic demonstration of a role for PKA in the late phase of LTP and in hippocampus-based long-term memory." *Cell* **88**: 615-626.

Alberini, C.M., M. Ghirardi, R. Metz, and E.R. Kandel (1994). "C/EBP is an immediate early gene required for the consolidation of long-term facilitation in *Aplysia*." *Cell* **76**: 1099-1114.

Albright, T.D., R. Desimone, and C. Gross (1994). "Columnar organization of directionally selective cells in visual area MT of macaques." *Journal of Neurophysiology* **51**: 16-31.

Arancio, O, M. Kiebler, C. Lee, V. Lev-Ram, R. Tsien, E.R. Kandel, and R. Hawkins (1996). "Nitric oxide acts directly in the presynaptic neuron to produce long-term potentiation in cultured hippocampal neurons." *Cell* **87**: 1025-1035.

Arancio, O., I. Antonova, S. Gambaryan, S. Lohmann, J. Wood, D. Lawrence, and R. Hawkins (2000). "Presynaptic role of cGMP-dependent protein kinase during long-lasting potentiation." *Journal of Neuroscience* **21**: 143-149.

Arthurs, O. and S. Boniface (2002). "How well do we understand the neural origins of the fMRI-BOLD signal?" *Trends in Neurosciences* **25**: 27-31.

Baars, B. (1997). *In the Theater of Consciousness*. Oxford: Oxford University Press.

Baddeley, A. (1986). *Working Memory*. Oxford: Clarendon Press.

Baddeley, A. (1994). "Working memory." In M. Gazzaniga (ed.), *The Cognitive Neurosciences*. Cambridge, MA: MIT Press, 755-764.

Bahr, B., U. Staubli, P. Xiao, D. Chun, Z. Ji., E. Esteban, and G. Lynch (1997). "Arg-Gly-Asp-Ser-Selective Adhesion and the stabilization of long-term potentiation: Pharmacological studies and the characterization of a candidate matrix receptor." *Journal of Neuroscience* **17**: 1320-1329.

Bailey, C., D. Bartsch, and E.R. Kandel (1996). "Toward a molecular definition of long-term potentiation." *Proceedings of the National Academy of Sciences* **93**: 13445-13452.

Bailey, C.H. and M. Chen, (1989). "Time course of structural changes at identified sensory neuron synapses during long-term sensitization in *Aplysia*." *Journal of Neuroscience* **9**: 1774-1780.

Baker, L.R. (1993). "Metaphysics and mental causation." In J. Heil and A. Mele (eds.), *op. cit.*

Balzer, W. and C.U. Moulines (1996). *Structuralist Theory of Science*. Walter de Gruyter, Berlin.

Balzer, W., C.U. Moulines, and J.D. Sneed (1987). *An Architectonic for Science*. Reidel, Dordrecht.

Bartsch, D., A. Casadio, K.A. Karl, P. Serodio, and E.R. Kandel (1998). "CREB1 encodes a nuclear activator, a repressor, and a cytoplasmic modulator that form a regulatory unit critical for long-term facilitation." *Cell* **95**: 211-223.

Bartsch, D., M. Ghirardi, A. Casadio, M. Giustetto, K.A. Karl, H. Zhu, and E.R. Kandel (2000). "Enhancement of memory-related long-term facilitation by ApAF, a novel transcription factor that acts downstream from both CREB1 and CREB2." *Cell* **103**: 595-608.

Bartsch, D., M. Ghirardi, P. Skehel, K. Karl, S. Herder, M. Chen, C. Bailey, and E.R. Kandel (1995). "*Aplysia* CREB2 represses long-term facilitation: Relief of repression converts transient facilitation into long-term functional and structural change." *Cell* **83**: 979-992.

Bechtel, W. and J. Mundale (1999). "Multiple realizability revisited: Linking cognitive and neural states." *Philosophy of Science* **66**: 175-207.

Bechtel, W. and R. Richardson (1993). *Discovering Complexity: Decomposition and Localization as Strategies in Scientific Research.* Princeton: Princeton University Press, 1993.

Beckermann, A. (2001). "Physicalism and new wave reductionism." *Grazer Philosophische Studien* **61**, 257-261.

Bergold. P., J. Sweatt, I. Winicov, K. Weiss, E.R. Kandel, and J.M. Schwartz (1990). "Protein synthesis during acquisition of long-term facilitation is needed for the persistent loss of regulatory subunits of the *Aplysia* cAMP-dependent protein kinase." *Proceedings of the National Academy of Sciences USA* **87**: 3788-3791.

Bernabeu, R., L. Bevilaqya, P. Ardenghi, E. Bromberg, P. Schmitz, M. Bianchin, I. Izquierdo, and J. Medina, J. (1997). "Involvement of hippocampal cAMP/cAMP-dependent protein kinase signaling pathways in a late memory consolidation phase of aversively motivated learning in rats." *Proceedings of the National Academy of Sciences* **94**: 7041-7046.

Bernstein, M., S. Stiehl, and J. Bickle (2000). "The effect of motivation on the stream of consciousness: Generalizing from a neurocomputational model of cingulo-frontal circuits controlling saccadic eye movements." In R. Ellis and N. Newton (eds.), *The Caldron of Consciousness: Motivation, Affect, and Self-Organization.* Amsterdam: John Benjamins, 133-161.

Bickle, J. (1989). *Towards a Scientific Reformulation of the Mind-Body Problem.* Doctoral dissertation, University of California, Irvine.

Bickle, J. (1992a). "Multiple realizability and psychophysical reduction." *Behavior and Philosophy* **20**: 47-58.

Bickle, J. (1992b). "Revisionary physicalism." *Biology and Philosophy* **7**: 411-430.

Bickle, J. (1993). "Connectionism, eliminativism, and the semantic view of theories." *Erkenntnis* **39**: 359-382.

Bickle, J. (1995a). "Connectionism, reduction, and multiple realizability." *Behavior and Philosophy* **23**: 29-39.

Bickle, J. (1995b). "Psychoneural reduction of the genuinely cognitive: Some accomplished facts." *Philosophical Psychology* **8**: 265-285.

Bickle, J. (1998). *Psychoneural Reduction: The New Wave.* Cambridge, MA: MIT Press.

Bickle, J. (1999). "Multiple realizability." In E. Zalta (ed.) *Stanford Encyclopedia of Philosophy* (http://plato.stanford.edu/entries/multiple-realizability/)

Bickle, J. (2000). "Concepts of intertheoretic reduction in current philosophy of mind." In *A Field Guide to Philosophy of Mind* (Societa Italiana Filosofia Analitica). http://www.uniroma3.it/kant/field/

Bickle, J. (2001). "New wave metascience: Replies to Beckerman, Maloney, and Stephen." *Grazer Philosophische Studien* **61**: 285-293.

Bickle, J. (2002). "Concepts structured through reduction: A structuralist resource illuminates the consolidation-long-term potentiation (LTP) link." *Synthese* **130**: 123-133.

Bickle, J., C. Worley, and M. Bernstein (2000). "Vector subtraction implemented neurally: A neurocomputational model of some sequential cognitive and conscious processes." *Consciousness and Cognition* **9**: 117-144.

Bickle, J., S. Holland, V. Schmithorst, and M. Avison (2001). "Cellular mechanisms of sequential processing in frontal regions revealed using a combined computational-fMRI methodology." *Proceedings of the First World Congress in Neuroinformatics*. Technical University Vienna: 581-597.

Bliss, T. and A. Gardner-Medwin (1973). "Long-lasting potentiation of synaptic transmission in the dentate area of the unanaesthetized rabbit following stimulation of the perforant path." *Journal of Physiology (London)* **232**: 357-374.

Bliss, T. and T. Lømo (1973). "Long-lasting potentiation of synaptic transmission in the dentate area of the anaesthetized rabbit following stimulation of the perforant path." *Journal ofPhysiology (London)* **232**: 331-356.

Block, N. ([1990] 1997). "Inverted earth." Reprinted in N. Block, O. Flanagan, and G. Guzeldere (eds.), *The Nature of Concsciousness*. Cambridge, MA: MIT Press, 677-693.

Block, N. (1978). "Troubles with functionalism." In C. Savage (ed.), *Perception and Cognition: Issues in the Foundations of Psychology. Minnesota Studies in the Philosophy of Science, volume 9*. Minneapolis: University of Minnesota Press, 261-325.

Bontley, T. (2000). "Review of Bickle, *Psychoneural Reduction: The New Wave*." *British Journal for the Philosophy of Science* **51**: 901-905.

Bourtchouladze, R., B. Frenguelli, J. Blendy, D. Cioffi, G. Schutz, and Silva, A. (1994). "Deficient long-term memory in mice with a targeted mutation of the cAMP-responsive element binding protein.' *Cell* **79**: 59-68.

Britten, K. and R. van Wezel (1998). "Electrical microstimulation of cortical area MST biases heading perception in monkeys." *Nature Neuroscience* **1**: 59-63.

Bruce, C. and M. Goldberg (1985). "Primate frontal eye fields. I. Single neurons discharging before saccades." *Journal of Neurophysiology* **53**: 603-635.

Bunsey, M. and H. Eichenbaum (1996). "Conservation of hippocampal memory function in rats and humans. *Nature* **379**: 255-257.

Burge, T. (1979). "Individualism and the mental." In P. French, T. Euhling, and H. Wettstein (eds.), *Midwest Studies in Philosophy, Volume 4: Studies in Epistemology*. Minneapolis: University of Minnesota Press.

Burge, T. (1993). "Mind-body causation and explanatory practice." In J. Heil and A. Mele (eds.), *op. cit.*

Burman, D. and C. Bruce (1997). "Suppression of task-related saccades by electrical stimulation in the primate's frontal eye fields." *Journal of Neurophysiology* **77**: 2252-2267.

Cammarota, M., L. Bevilaqua, P. Ardenghi, G. Paratcha, M. Levi de Stein, I. Izquierdo, and J. Medina (2000). "Learning-associated activation of nuclear MAPK, CREB and Elk-1, along with fos production, in the rat hippocampus after one-trial avoidance learning: Abolition by NMDA receptor blockade." *Molecular Brain Research* **76**: 36-46.

Campbell, N. and J. Reece (2001). *Biology*, 6th Ed. New York: Benjamin/Cummings.

Carlson, N.R. (1994). *Physiology of Behavior*, 5th Ed. Boston: Allyn and Bacon.

Carnap, R. ([1932] 1958). "The elimination of metaphysics through logical analysis of language" (trans. Arthur Pap). Reprinted in A.J. Ayer (ed.), *Logical Positivism*. Glencoe, IL: Free Press, 1959, 60-81.

Carnap, R. ([1950], 1956). "Empiricism, semantics, ontology." Reprinted in R. Carnap, *Meaning and Necessity*, 2nd. Ed. Chicago: University of Chicago Press, 205-221.

Carnap, R. (1934). *The Unity of Science* (trans. M. Black). London: Kegan, Paul, Trench, Trubner and Company.

Chafee, M. and P.S. Goldman-Rakic (1998). "Matching patterns of activity in primate prefrontal area 8a and and parietal area 7ip neurons during a spatial working memory task." *Journal of Neurophysiology* **79**: 2919-2940.

Chain, D., A. Casadio, S. Schacher, A. Hegde, M. Valbrun, N. Yamamoto, A. Goldberg, D. Bartsch, E.R. Kandel, and J. Schwartz (1999). "Mechanisms for generating the autonomous cAMP-dependent protein kinase required for long-term facilitation in *Aplysia*.' *Neuron* **22**: 147-156.

Chalmers, D. (1995). "Facing up to the problem of consciousness." *Journal of Consciousness Studies* **2**: 200-219.

Chalmers. D. (1996). *The Conscious Mind*. Oxford: Oxford University Press.

Churchland, P.M. (1982). "Is *thinker* a natural kind?" *Dialogue* **21**: 223-238.

Churchland, P.M. (1985). "Reduction, qualia, and the direct introspection of brain states." *Journal of Philosophy* **82**: 1-22.

Churchland, P.M. (1987). *Matter and Consciousness*, Revised Ed. Cambridge, MA: MIT Press.

Churchland, P.M. (1995). *The Engine of Reason, The Seat of the Soul*. Cambridge, MA: MIT Press.

Churchland, P.S. (1986). *Neurophilosophy: Towards a Unified Science of the Mind/Brain*. Cambridge, MA: MIT Press.

Churchland, P.S. and T.J. Sejnowski (1992). *The Computational Brain*. Cambridge, MA: MIT Press.

Clayton, D.F. (2000). "The genomic action potential." *Neurobiology of Learning and Memory* **74**: 185-216.

Cohen, N. and H. Eichenbaum (1993). *Memory, Amnesia, and the Hippocampal System*. Cambridge, MA: MIT Press.

Compte, A., N. Brunel, P.S. Goldman-Rakic, and X. Wang (2000). "Synaptic mechanisms and network dynamics underlying spatial working memory in a cortical network model." *Cerebral Cortex* **10**: 910-923.

Connolly, J.B. and T. Tully (1997). "Behaviour, learning and memory." In D. Roberts (ed.), *Drosophila: A Practical Approach*. Oxford: Oxford University Press.

Courtney, S., Petit, L., Maisong, J., Ungerleider, L., and Haxby, J. (1998). "An area specialized for spatial working memory in the human frontal cortex." *Science* **279**: 1347-1351.

Craver, C. (2003). "The making of a memory mechanism." Forthcoming in *Journal of the History of Biology*.

Craver, C. and L. Darden (2001). "Discovering mechanisms in neurobiology: The case of spatial memory." In P. K. Machamer, R. Grush, and P. McLaughlin (eds.), *Theory and Method in Neuroscience*. Pittsburgh, PA: University of Pittsburgh Press.

Davis, H. and L. Squire (1984). "Protein synthesis and memory: A review." *Psychological Bulletin* **96**: 518-559.

DeAngelis, G., B. Cumming, and W. Newsome (1998). "Cortical area MT and the perception of stereoscopic depth." *Nature* **394**: 677-680.

Dennett, D.C. (1991). *Consciousness Explained*. Boston: Little Brown.

Downing, C. (1988). "Expectancy and visual-spatial attention: Effects on perceptual quality." *Journal of Experimental Psychology: Human Perception and Performance* **14**: 188-202.

Dretske, F. (1996). "Phenomenal externalism: If meanings ain't in the head, where are qualia?" In E. Villaneuva (ed.), *Philosophical Issues* **7**. Atascadero, CA: Ridgeview, 143-158.

Dubnau, J. and T. Tully (1998). "Gene discovery in *Drosophila*: New insights for learning and memory." *Annual Review of Neuroscience* **21**: 407-444.

Dudai, Y. and W.G. Quinn (1976). "Memory phases in *Drosophila*." *Nature* **262**: 576-577.

Duncan, C. (1949). "The retroactive effect of electroshock on learning." *Journal of Comparative and Physiological Psychology* **42**: 32-44.

Eichenbaum, H. (1997). "Declarative memory: Insights from cognitive neurobiology." *Annual Review of Psychology* **48**: 547-572.

Enç, B. (1983). "In defense of the identity theory." *Journal of Philosophy* **80**: 279-298.

Endicott, R (2001). "Post-structuralist angst: Critical notice of Bickle's *Psychoneural Reduction: The New Wave*." *Philosophy of Science* **68**: 377-393.

Feigl, H. (1967). *The 'Mental' and the 'Physical': The Essay and a Postscript*. Minneapolis: University of Minnesota Press.

Feyerabend, P.K. (1962). "Explanation, reduction, and empiricism." *Minnesota Studies in the Philosophy of Science*, vol. 3. Reprinted in P.K. Feyerabend, *Realism, Ratonalism and Scientific Method: Philosophical Papers*, vol. 1. Cambridge: Cambridge University Press, 44-96.

Fodor, J. (1975). *The Language of Thought*. New York: Thomas Crowell.

Fodor, J.A. ([1974] 1981). "Special sciences." Reprinted in J.Fodor, *RePresentations*. Cambridge, MA: MIT Press, 127-145..

Freeman, W. (2000). *Neurodynamics: An Exploration in Mesoscopic Brain Dynamics*. New York: Springer.

Frost, W.N., V.F. Castellucci, R.D. Hawkins, and E.R. Kandel (1985). "Monosynaptic connections from the sensory neurons of the gill- and siphon-withdrawal reflex in *Aplysia* participate in the storage of long-term memory for sensitization." *Proceedings of the National Academy of Sciences USA* **82**: 8266-8269.

Funahashi, S., C. Bruce, and P.S. Goldman-Rakic (1989). "Mnemonic coding of visual space in the monkey's dorsolateral prefrontal cortex." *Journal of Neurophysiology* **61**: 1-19.

Funahashi, S., C. Bruce, and P.S. Goldman-Rakic (1991). "Neuronal activity related to saccadic eye movements in the monkey's dorsolateral prefrontal cortex." *Journal of Neurophysiology* **65**:1464-1483.

Funahashi, S., M. Chafee, and P.S. Goldman-Rakic (1993). "Prefrontal neuronal activity in rhesus monkeys performing an anti-saccade task." *Nature* **365**: 753-756.

Garcia, J. and R. Koelling (1966). "Relation of cue to consequence in avoidance learning." *Psychonomic Science* **4**: 123-124.

Gaymard, B., S. Rivaud, J. Cassarini, T. Dubard, G. Rancurel, Y. Agid, and C. Pierrot-Deseilligny, C. (1998). "Effects of anterior cingulate lesions on ocular saccades in humans." *Experimental Brain Research* **120**: 173-183.

Gilbert, C., M. Ito, M. Kapadia, and G. Westheimer (2000). "Interactions between attention, context and learning in primary visual cortex." *Vision Research* **40**: 1217-1226.

Goldberg, M. and Bruce, C. (1990). "Primate frontal eye fields. III. Maintenance of spatially accurate saccade signal." *Journal of Physiology* **64**: 489-508.

Goldberg, M., H. Eggers, and P. Gouras (1991). "The oculomotor system." In E.R. Kandel, J. Schwartz, and T. Jessell (eds.), *op. cit.*

Goldman-Rakic, P.S. (1995). "Cellular basis of working memory." *Neuron* **14**: 477-485.

Goldman-Rakic, P.S. (1996). "Regional and cellular fractionation of working memory." *Proceedings of the National Academy of Sciences (USA)* **93**: 13473-13480.

Goldman-Rakic, P.S., S. O'Scalaidhe, and M. Chafee (2000). "Domain specificity in cognitive systems." In M. Gazzaniga (ed.), *The New Cognitive Neurosciences*. Cambridge, MA: MIT Press, 733-742.

Goodale, M. and A. Milner (1992). "Separate visual pathways for perception and action." *Trends in Neurosciences* **15**. 20-25.

Hannan, B. (2000). "Review of Bickle, *Psychoneural Reduction: The New Wave*." *Philosophical Books* **41**: 53-54.

Hardcastle, V.G. (1995). *Locating Consciousness*. Amsterdam: John Benjamins.

Harman, G. (1982). "Conceptual role semantics." *Notre Dame Journal of Formal Logic* **23**.

Haus-Seuffert, P. and M. Meisterernst (2000). "Mechanisms of transcriptional activation of cAMP-responsive element-binding protein CREB." *Molecular and Cellular Biochemistry* **212**: 5-9.

Hawkins, R.D. and E.R. Kandel (1984a). "Is there a cell-biological alphabet for simple forms of learning?" *Psychological Review* **91**: 375-391.

Hawkins, R.D. and E.R. Kandel (1984b). "Steps toward a cell-biological alphabet for elementary forms of learning." In G. Lynch, J.L. McGaugh, and N.M. Weinberger (eds.), *Neurobiology of Learning and Memory*. New York: Guilford Press, 385-404.

Hebb, D. 1949: *The Organization of Behavior*. John Wiley: New York.

Hegde, A.N., A.L. Goldberg, and J.H. Schwartz (1993). "Regulatory subunits of cAMP-dependent protein kinases are degraded after conjugation to ubiquitin: A molecular mechanism underlying long-term synaptic plasticity." *Proceedings of the National Academy of Sciences USA* **90**: 7436-7440.

Hegde, A.N., K. Inokuchi, W. Pei, A. Casadio, M. Ghirardi, D.G. Chain, K.C. Martin, E.R. Kandel, and J.H. Schwartz (1997). "Ubiquitin C-Terminal Hydrolase is an immediate early gene essential for long-term facilitation in *Aplysia*." *Cell* **89**: 115-126.

Heil, J. and A.Mele (eds.) (1993). *Mental Causation*. Oxford: Clarendon Press.

Holland, S., E. Plante, A. Weber-Byers, R. Strawsburg, V. Schmithorst, and W. Ball (2001). "Normal fMRI brain activation patterns in children performing a verb generation task." *Neuroimage* **14**: 837-843.

Hooker, C.A. (1981). "Towards a general theory of reduction. Part I: Historical and scientific setting. Part II: Identity in reduction. Part III: Cross-categorial reduction." *Dialogue* **20**: 38-59, 201-236, 496-529.

Horgan, T. (1993). "Nonreductive materialism and the explanatory autonomy of psychology," In S. Wagner and R. Warner (eds.), *Naturalism: A Critical Appraisal*. Notre Dame, IN: University of Notre Dame Press, 295-320.

Horgan, T. (2001). "Causal compatibilism and the exclusion problem." *Theoria* **16**: 95-116.

Hummler, E., T. Cole, J. Blendy, R. Ganss, A. Aguzzi, W. Schmid, F. Beermann, and G. Schutz (1994). "Targeted mutation of the CREB gene: Compensation within the CREB/ATF family of transcription factors." *Proceedings of the National Academy of Sciences (USA)* **91**: 5647-5651.

Izquierdo, I. and J. Medina (1997). "Memory formation: The sequence of biochemical events in the hippocampus and its connection to activity in other brain structures.' *Neurobiology of Learning and Memory* **68**: 285-316.

Jackson, F. (1983). "Epiphenomenal qualia." *Philosophical Quarterly* **32**: 127-136.

James, W. (1890). *The Principles of Psychology*. New York: Henry Holt.

Kandel, E.R. (1979). "Cellular insights into behavior and learning." *Harvey Lectures* **73**: 19-92.

Kandel, E.R., J.R. Schwartz, and T. Jessell (eds.) (1991). *Principles of Neural Science*, 3rd Ed. New York: McGraw-Hill.

Kandel, E.R., J.R. Schwartz, and T. Jessell (eds.) (2000). *Principles of Neural Science*, 4th Ed. New York: McGraw-Hill.

Kim, J. (1990): 'Explanatory exclusion and the problem of mental causation.' In E. Villanueva (ed.), *Information, Semantics, and Epistemology*. Basil Blackwell: Oxford, 36-56.

Kim, J. (1993). *Supervenience and Mind*. Cambridge: Cambridge University Press.

Kim, J. (1996). *Philosophy of Mind*. Boulder, CO: Westview.

Kim, J. (1998). *Mind in a Physical World*. Cambridge, MA: MIT Press.

Kimura, M. (1983). *The Neutral Theory of Molecular Evolution*. Cambridge: Cambridge University Press.

Kolb, B. and I. Whishaw (1996). *Fundamentals of Human Neuropsychology*, 4th Ed. W.H. Freeman: New York.

Kosslyn, S. (1997). "Mental imagery." In M. Gazzaniga (ed.), *Conversations in the Cognitive Neurosciences*. Cambridge, MA: MIT Press, 155-174.

Kritzer, M. and P.S. Goldman-Rakic (1995). "Intrinsic circuit organization of the major layers and sublayers of the dorsolateral prefrontal cortex of the rhesus monkey." *Journal of Comparative Neurology* **359**: 131-143.

Kuhn, T. (1962). *The Structure of Scientific Revolutions*. Chicago: University of Chicago Press.

LeDoux, J.E. (2000). "Emotion circuits in the brain." *Annual Review of Neuroscience* **23**: 155-184.

LePore, E. and B. Loewer (1989). "More on making mind matter." *Philosophical Topics* **17**: 175-191.

Levin, L., P. Han, P. Hwang, P. Feinstein, R. Davis, and R. Reed (1992). "The *Drosophila* learning and memory gene *rutabaga* encodes a Ca^{2+}/calmodulin-responsive adenylyl cyclase." *Cell* **68**: 479-489.

Levine, J. (1983). Materialism and qualia: The explanatory gap." *Pacific Philosophical Quarterly* **64**: 354-361.

Levitan, I. And L. Kaczmarek (2001). *The Neuron, 3rd Ed.* Oxford: Oxford University Press.

Lewin, B. 1999: *Genes VII.* Oxford University Press: Oxford.

Lewis, D. (1969). "Review of *Art, Mind, and Religion.*" *Journal of Philosophy* **66**: 23-35.

Lewis, D. (1973). "Scorekeeping in a Language Game." *Journal of Philosophical Logic* **8**: 339-59.

Liu, J. and W. Newsome (2000). "Somatosensation: Touching the mind's fingers." *Current Biology* **10**: R598-R600.

Llinás, R. and P.S. Churchland (1996). "Introduction." In R. Llinás and P.S. Churchland (eds.), *The Mind-Brain Continuum.* Cambridge, MA: MIT Press, ix-xi.

Lodish, H., A. Berk, L. Zipursky, P. Matsudaira, D. Baltimore, and J. Darnell (2000). *Molecular Cell Biology*, 4th Ed. New York: W.H. Freeman.

Loughland, C. L. Williams, and E. Gordon (2002). "Visual scanpaths to positive and negative facial emotions in an outpatient schizophrenia sample." *Schizophrenia Research* **55**: 159-170.

Lycan, W. (1996). *Consciousness and Experience.* Cambridge, MA: MIT Press.

Lycan, W. (2001). "Response to Polger and Flanagan." *Minds and Machines* **11**: 127-132.

Lynch, G. (2000). "Memory consolidation and long-term potentiation." In M. Gazzaniga (ed.), *The New Cognitive Neurosciences.* Cambridge, MA: MIT Press, 139-157.

Lynch, G. and R. Granger (1989). "Simulation and analysis of a simple cortical network." *Psychology of Learning and Motivation* **23**: 205-241.

Martin, K., C. Casadio, Y. Zhu, J. Rose, M. Chen, C. Bailey, E.R. Kandel (1997). "Synapse-specific long-term facilitation of *Aplysia* sensory to motor synapses: A function for local protein synthesis in memory storage." *Cell* **91**: 927-938.

Maunsell, J. (1995). "The brain's visual world: Representation of visual targets in cerebral cortex." *Science* **270**: 764-769.

Mayford, M. and E.R. Kandel (1999). "Genetic approaches to memory storage." *Trends in Genetics* **15**: 463-470.

Mayr, D. (1976). "Investigations of the concept of reduction, I." *Erkenntnis* **10**: 275-294.

McAdams, C. and J. Maunsell (1999). "Effects of attention on orientation-tuning functions of single neurons in macaque cortical area V4." *Journal of Neuroscience* **19**: 431-441.

McAdams, C. and J. Maunsell (2000). "Attention to both space and feature modulates neuronal responses in macaque area V4." *Journal of Neurophysiology* **83**: 1751-1755.

McEachern, J. and C. Shaw (1996). "An alternative to the LTP orthodoxy: A plasticity-pathology continuum model." *Brain Research Reviews* **22**: 51-92.

McGaugh, J. (2000). "Memory: A century of consolidation.' *Science* **287**: 248-251.

McGinn, C. (1989). "Can we solve the mind-body problem?" *Mind* **98**: 349-366.

Montarolo, P., P. Goelet, V. Castellucci, J. Morgan, E.R. Kandel, and S. Schacher (1986). "A critical period for macromolecular synthesis in long-term heterosynaptic facilitation in *Aplysia*." *Science* **234**: 1249-1254.

Moulines, C.U. (1984). "Ontological reduction in the natural sciences." In W. Balzer, D. Pearce, and H.J. Schmidt (eds.), *Reduction in Science*. Dordrecht: Reidel, 51-70.

Mountcastle, V., M. Steinmetz, and R. Romo (1990). "Frequency discrimination in the sense of flutter: Psychophysical measurements correlated with postcentral events in behaving monkeys." *Journal of Neuroscience* **10**: 3032-3044.

Murasugi, C., C.D. Salzman, and W. Newsome (1993). "Microstimulation in visual area MT: Effects of varying pulse amplitude and frequency." *Journal of Neuroscience* **13**: 1719-1729.

Nagel, E. (1961). *The Structure of Science*. New York: Harcourt, Brace, and World.

Nagel, T. (1989). *The View from Nowhere*. Oxford: Oxford University Press,

Newsome, W. (1997). "Perceptual processes." In M. Gazzaniga (ed.), *Conversations in the Cognitive Neurosciences*. Cambridge, MA: MIT Press, 53-68.

O'Scalaidhe, S., F. Wilson, and P.S. Goldman-Rakic (1997). "Area segregation of face-processing neurons in prefrontal cortex." *Science* **278**: 1135-1138.

Pardo, J., P. Pardo, and M.E. Raichle (1993). "Neural correlates of self-induced dysphoria." *American Journal of Psychiatry* **150**:713-719.

Parent, A. (1996). *Carpenter's Human Neuroanatomy*, 9th Ed. Baltimore: Williams and Wilkins.

Penfield, W. and P. Perot (1963). "The brain's record of auditory and visual experience: A final summary and discussion." *Brain* **86**: 595-696.

Penrose, R. (1994). *Shadows of the Mind*. Oxford: Oxford University Press.

Petri, H. and M. Mishkin (1994). "Behaviorism, cognitivism, and the neuropsychology of memory." *American Scientist* **82**: 30-37.

Polger, T. and O. Flanagan (2001). "A decade of teleofunctionalism: Lycan's *Consciousness* and *Consciousness and Experience*." *Minds and Machines* **11**: 113-126.

Popper, K. (1962). "Truth, rationality, and the growth of scientific knowledge." In K. Popper, *Conjectures and Refutations*. New York: Basic Books, 215-250.

Posner, M. (1980). "Orientation of attention." *Quarterly Journal of Experimental Psychology* **32**: 3-25.

Posner, M. and M. Raichle (1994). *Images of Mind*. New York: Scientific American Library.

Prinz, J. (2000). "A neurofunctional theory of visual consciousness." *Consciousness and Cognition* **9**: 243-259.

Purves, W., G. Orians, H. Heller, and D. Sadava (1998). *Life: The Science of Biology*, 5th Ed. Sunderland, MA: Sinauer Associates.

Putnam, H. (1967). "Psychological predicates." In W. Capitan and D. Merrill (eds.), *Art, Mind, and Religion*. Pittsburgh: University of Pittsburgh Press. Reprinted in Putnam (1975).

Putnam, H. (1975a). *Mind, Language, and Reality: Philosophical Papers,* Vol. 2. Cambridge: Cambridge University Press.

Putnam, H. (1975b). "The meaning of 'meaning'." In K. Gunderson (ed.), *Language, Mind, and Knowledge: Minnesota Studies in the Philosophy of Science, volume 7*. Minneapolis: University of Minnesota Press.

Quine, W.V.O. ([1949] 1953). "Two dogmas of empiricism." Reprinted in W.V.O. Quine, *From a Logical Point of View*. New York: Harper and Row, 20-46.

Quine, W.V.O. ([1951] 1966). "On Carnap's views on ontology." Reprinted in W.V.O. Quine, *The Ways of Paradox and Other Essays*. New York: Random House, 126-134.

Quinn, W., W. Harris, and S. Benzer (1974). "Conditioned behavior in *Drosophila melanogaster*." *Proceedings of the National Academy of Sciences USA* **71**: 707-712.

Reichenbach, H. (1957). *The Rise of Scientific Philosophy*. Berkeley, CA: University of California Press.

Reichling, D. and J. Levine (1999). "The primary afferent nociceptor as pattern generator." *Pain* **6** (Supplement): S103-S109.

Richardson, R.C. (1979). "Functionalism and reduction." *Philosophy of Science* **46**: 533-558.

Richardson, R.C. (1999). "Cognitive science and neuroscience: New wave reductionism." *Philosophical Psychology* **12**: 297-307.

Ridley, M. (1996). *Evolution*, 2nd Ed. Cambridge, MA: Blackwell Science.

Romo, R., A. Hernández, A. Zainos, and E. Salinas (1998). "Somatosensory discrimination based on microstimulation." *Nature* **392**: 387-390.

Romo, R., A. Hernández, A. Zainos, C. Brody, and L. Lemus (2000). "Sensing without touching: Psychophysical performance based on cortical microstimulation." *Neuron* **26**: 273-278.

Rossi, A. and M. Paradiso (1995). "Feature-specific effects of selective visual attention." *Vision Research* **35**: 621-634.

Salzman, C.D., C. Murasugi, K. Britten, and W. Newsome (1992). "Microstimulation in visual area MT: Effects on direction discrimination performance." *Journal of Neuroscience* **12**: 2331-2355.

Saucier, D. and D. Cain (1995). "Spatial learning without NMDA receptor-dependent long-term potentiation." *Nature* **378**: 186-189.

Sawaguchi, T. and P.S. Goldman-Rakic (1994). "The role of D1-dopamine receptor in working memory: Local injections of dopamine antagonists into the prefrontal cortex of rhesus monkeys performing an oculomotor delayed-response task." *Journal of Neurophysiology* **71**: 515-528.

Schacter, D.L., C. Chu, and K. Ochsner (1993). "Implicit memory: A selective review." *Annual Review of Neuroscience* **16**: 159-183.

Schaffner, K. (1967). "Approaches to reduction." *Philosophy of Science* **34**: 137-147.

Schaffner, K. (1992). "Philosophy of medicine." In M. Salmon, J. Earman, C. Glymour, J. Lennox, P. Machamer, J. McGuire, J. Norton, W. Salmon, and K. Schaffner, *Introduction to the Philosophy of Science*. Englewood Cliffs, NJ: Prentice-Hall, 310-344.

Schmithorst, V. and B. Dardzinski (2000). "CCHIPS/IDL enables detailed MRI analysis." http://www.researchsystems.com/AppProfile/idl_med_cchips.cfm.

Schouten, M. and H. de Jong (1999). "Reduction, elimination, and levels: The case of the LTP-learning link." *Philosophical Psychology* **12**: 237-262.

Scoville, W. and B. Milner (1957). "Loss of recent memory after bilateral hippocampal lesions." *Journal of Neurology, Neurosurgery, and Psychiatry* **20**: 11-21.

Shepherd, G. (1994). *Neurobiology*, 3rd Ed. Oxford: Oxford University Press.

Smart, J.J.C. (1959). "Sensations and brain processes." *Philosophical Review* **68**: 141-156.

Smart, J.J.C. (1963). "Materialism." *Journal of Philosophy* **60**: 651-662.

Smart, J.J.C. (1967). "Comments on the papers." In C. Presley (ed.), *The Identity Theory of Mind*. St. Lucia, Brisbane: University of Queensland Press.

Sneed, J. (1971). *The Logical Structure of Mathematical Physics*. Dordrecht: Reidel.

Soderling, T. and V. Derkach (2000). "Postsynaptic protein phosphorylation and LTP." *Trends in Neuroscience* **23**: 75-80.

Son, H., R. Hawkins, K. Martin, M. Kiebler, P. Huang, M. Fishman, and E.R. Kandel (1996). "Long-term potentiation is reduced in mice that are doubly mutant in endothelial and neuronal nitric oxide synthase.' *Cell* **87**: 1015-1023.

Squire, L. (1992). "Memory and the hippocampus: A synthesis from findings with rats, monkeys, and humans." *Psychological Review* **99**: 195-231.

Squire, L. 1987: *Memory and Brain*. Oxford University Press, Oxford.

Squire, L. and E.R. Kandel (1999). *Memory: From Mind to Molecules*. Scientific American Library, New York.

Staubli, U., D. Chun, and G. Lynch (1998). "Time-dependent reversal of long-term potentiation by an integrin antagonist." *Journal of Neuroscience* **18**: 3460-3469.

Stegmüller, W. (1976). *The Structure and Dynamics of Theories*. Berlin: Springer-Verlag.

Stephen, A. (2001). "How to lose the mind-body problem." *Grazer Philosophische Studien* **61**: 279-283.

Suppes, P. (1956). *Introduction to Logic*. Princeton, NJ: van Nostrand.

Suppes, P. (1965). "What is a scientific theory?" In S. Morgenbesser (ed.), *Philosophy of Science Today*. New York: Basic Books.

Taubenfield, S., K. Wiig, B. Monti, B. Dolan, G. Pollonini, and C. Alberini (2001). "Fornix-dependent induction of hippocampal CCAAT enhancer-binding protein β and δ co-localizes with phosphorylated cAMP response element-binding protein and accompanies long-term memory consolidation.' *Journal of Neuroscience* **21**: 84-91.

Tully, T. and W. Quinn (1985). "Classical conditioning and retention in normal and mutant *Drosophila melanogaster*." *Journal of Comparative Physiology A* **157**: 263-277.

Tulving, E. (1983). *Elements of Episodic Memory*. New York: Oxford University Press.

Ungerleider, L. and M. Mishkin (1982). "Two cortical visual systems." In D. Ingle, M. Goodale, and R. Mansfield (eds.), *Analysis of Visual Behavior*. Cambridge, MA: MIT Press, 549-586.

Weaver, R. (1999). *Molecular Biology*, Boston, MA: WCB/McGraw-Hill.

Williams, G. and P.S. Goldman-Rakic (1995). "Modulation of memory fields by dopamine D1 receptors in prefrontal cortex." *Nature* **376**: 572-575.

Wilson, F., S. O'Scalaidhe, and P.S. Goldman-Rakic (1993). "Dissociation of object and spatial processing domains in primate prefrontal cortex." *Science* **260**: 1955-1958.

Wilson, F., S. O'Scalaidhe, and P.S. Goldman-Rakic (1994). "Functional synergism between putative gamma-aminobutyrate-containing neurons and pyramidal neurons in prefrontal cortex." *Proceedings of the National Academy of Sciences (USA)* **91**: 4009-4013.

Wimsatt, W.C. (1974). "Reductive explanation: A functional account." In R.S. Cohen *et al.* (eds.), *Philosophy of Science Association Proceedings*. Dordrecht: Reidel.

Wittgenstein, L. ([1919]1961). *Tractatus Logico-Philosophicus* (translated by D. Pears and B. McGuiness). London: Routledge and Kegan Paul.

Yin, J., J. Wallach, E. Wilder, J. Klingensmith, D. Dang, N. Perrimon, H, Zhou, T. Tully, and W. Quinn (1995b). "A *Drosophila* CREB/CREM homolog encodes multiple isoforms, including a cyclic AMP-dependent protein kinase-responsive transcriptional activator and antagonist." *Molecular and Cellular Biology* **15**: 5123-5130.

Yin, J., J. Wallach, M. Del Vecchio, E. Wilder, H. Zhou, W. Quinn, and T. Tully (1994). "Induction of a dominant negative CREB transgene specifically blocks long-term memory in *Drosophila*." *Cell* **79**: 49-58.

Yin, J., M. Del Vecchio, H. Zhou, and T. Tully (1995a). "CREB as a memory modulator: Induced expression of a *dCREB2* isoform enhances long-term memory in *Drosophila*." *Cell* **81**: 107-115.

Zangwell, N. (1992). "Variable reduction not proven." *Philosophical Quarterly* **42**: 214-218.

Zigmond, M., F. Bloom, S. Landis, J. Roberts, and L. Squire (eds.) (1999). *Fundamental Neuroscience*. San Diego, CA: Academic Press.

INDEX